Meteorology and Geography

NIPA® GENX ELECTRONIC RESOURCES & SOLUTIONS P. LTD.
New Delhi-110 034

About the Authors

Surender Singh has about thirty five years of experience with PhD in the field of Agrometeorology. Worked in Agrometeorology encompassing teaching, research, extension and administration at various capacities viz., Principal Scientist, Professor & Head, Senior Scientist, Assistant Scientist, Advisor (Recruitments) and Associate Director (Students' Counseling and Placement) at CCS Haryana Agricultural University, Hisar, India. Recipient of Two Gold Medals for Best Research Work during PhD Program and for Best PhD Dissertation (2001-02) awarded by then President of India Dr APJ Abdul Kalam. Offered UK Universities Vice Chancellors' Scholarship for Higher Studies twice (1991 & 1993) under Commonwealth Fellowship Program. Post Doctoral Fellow at Abdus Salam International Centre for Theoretical Physics, Trieste, Italy; IAC – Instituto Agronômico, Campinas, Brazil and BOKU (Met) - Institute of Meteorology and Climatology, Vienna, Austria.

He is life member of Association of Agrometeorologists, Indian Meteorological Society, Indian Society of Remote Sensing, International Society for Agrometeorology, Society for Sustainable Agric& Resource Management, Environment Science Published for Everybody Round the Earth (ESPERE), European Drought Centre, South Asian Forum on Agriculture Meteorology (SAFOAM) and The Geographical Society of India. Organized quite a few trainings/symposia in Agrometeorology and associated domains at National level. Published more than 250 research papers, popular articles, manuals/teaching aids and research bulletins/book chapters etc. Performed responsibilities as Managing Editor - Journal of Agrometeorology; Research Collaborator, Monsoon Asia Agro-Environmental Research Consortium (MARCO), Japan; SPARC Research Collaborator, ETH, Zurich, Switzerland; Member, Expert Team - CAgM-WMO, Geneva, Switzerland (2018-20). Core Committee Member on Agriculture – Drought at WMO, Geneva, Switzerland (2021-24 & 2024-27). External Expert for evaluation of research projects, COST-Association, European Union, Brussels, Belgium. Many students (UG, PG and PhD) have been trained under him and placed across the country and well settled in various positions in the field of Agrometeorology. He is widely travelled researcher/expert across the globe on many academic pursuits. He has been conferred upon Fellow of Association of Agrometeorologists (FAAM) for the year 2024 in recognition of his exemplary achievements and enduring impact on agrometeorological science.

V Uma Maheswara Rao has forty years of experience with PhD in the broader field of Agrometeorology. Worked in Agrometeorology encompassing teaching, research, extension and administration at various capacities viz., Project Coordinator, Principal Scientist, Professor & Head, Agrometeorologist, and Assistant Scientist at CCS Haryana Agricultural University and Indian Council of Agriculture Research in India.

He is life member of Indian Meteorological Society, The Arid Zone Research Association of India, Indian Society of Remote Sensing, Association of Agrometeorologists, International Society for Agrometeorology, and The Indian Society of Dryland Agriculture. Organized several trainings/workshops in Agrometeorology at the International and National level. Published more than 345 research papers, popular articles, manuals/teaching aids and research bulletins/book chapters. Member of the CCl-ET-IIC, WMO, Geneva (2014-2017). Member of XI Plan High Powered Review Committee of Ministry of Earth Sciences, Government of India (2012). In 2014, the prestigious Chaudhary Devi Lal Award, ICAR's Best AICRP on Agrometeorology was received under his dynamic leadership as Project Coordinator. Many students (UG, PG and PhD) have been trained under his guidance are placed across the country and settled in various positions in the field of Agrometeorology. He visited numerous countries in official capacity.

BV Ramana Rao taught Agricultural Meteorology in the University of Agricultural Sciences, Bangalore during the years 1966 to 1979 and later joined the Indian Council of Agricultural Research and superannuated from service during April, 2000. He served as Head, Division of Climatology, Wind Power and Solar Energy Utilization at ICAR-Central Arid Zone Research Institute, Jodhpur (Rajasthan); founder Project Coordinator, All India Coordinated Research Project on Agrometeorology at ICAR Central Research Institute for Dryland Agriculture, Hyderabad (Telangana) during December 1984 to March 1994. His major contributions in the field of Agrometeorology include Agroclimatic Characterization, Delineation of Efficient Cropping Zones, Crop Weather Modeling, Classification and Management of Agricultural Droughts, Operational Agrometeorology, etc. He published about 150 research papers in peer reviewed journals with 50 of his papers appearing in high impact International Journals. He authored 15 books and monographs and a number of scientific and technical bulletins. He served as Editor in Chief, Journal of Agrometeorology during the years 2009 to 2023.

He is the recipient of British Council Fellowship during the year 1986-87 for advanced training at University of Reading, United Kingdom. He was conferred as Honorary Fellow (2011) and Life Time Achievement Award (2025) by the Association of Agrometeorologists for his valuable contribution to the growth and development of Agricultural Meteorology in India.

Meteorology and Geography

Surender Singh
MSc; PhD; FAAM
Professor/Principal Scientist (Agrometeorology)
CCS HAU, Hisar, Haryana, India

V Uma Maheswara Rao
MSc; PGDIAS; PhD; FAAM
Former Project Coordinator - AICRP on Agrometeorology
ICAR-CRIDA, Hyderabad, Telangana, India

BV Ramana Rao
MSc Tech; FAAM
Founder Project Coordinator - AICRP on Agrometeorology
ICAR-CRIDA, Hyderabad, Telangana, India

nipa®

NIPA® GENX ELECTRONIC RESOURCES & SOLUTIONS P. LTD.
New Delhi-110 034

NIPA® GENX ELECTRONIC RESOURCES & SOLUTIONS P. LTD.

101,103, Vikas Surya Plaza, CU Block
L.S.C.Market, Pitam Pura, New Delhi-110 034
Ph : +91 11 4386 0225, 9717133558, 9540816132
E-mail: newindiapublishingagency@gmail.com
Website: www.niparesources.com

Print ISBN: 978-93-72197-01-3

ebook ISBN: 978-93-72193-43-5

NIPA® also publishes books in a variety of electronic formats. Some content that appears in print may not be available in electronic books, and vice versa

Composed and Designed by NIPA®.

The Book

Meteorology and Geography

is dedicated to

Dr M Ravichandran

(b May 18, 1965)

This **book, 'Meteorology & Geography'** is dedicated to **Dr Muthalagu Ravichandran,** Secretary, MoES, Govt of India in gratitude and admiration for his transformative leadership. Dr Ravichandran- A visionary leader in Earth system sciences and an ardent advocate for ocean-atmosphere research integration, whose unwavering commitment to advancing meteorological and geographical knowledge has inspired a generation of scientists and academicians.

His visionary guidance has been instrumental in shaping Mission Mausam into a beacon of innovation—advancing how we forecast, interprets, and applies climate information to support India's various sectors including farming communities. His pioneering work in polar and oceanographic studies, coupled with his leadership in national and international scientific initiatives, has significantly enriched our understanding of Earth's complex systems.

We dedicate this book to him with deep respect and gratitude, in recognition of his distinguished service, scientific foresight, and enduring contribution to climate and geospatial sciences in India and beyond.

Dr M Mohapatra

Director General of Meteorology
India Meteorological Department
Ministry of Earth Sciences
Government of India, New Delhi

Foreword

The intricate relationship between our atmosphere, landmasses, and the aquatic world forms the very bedrock of sustainable fisheries. The disciplines of meteorology and geography are not peripheral but central to understanding the dynamics of our oceans, rivers, and lakes. From the subtle shifts in atmospheric pressure that influence ocean currents and upwelling, to the patterns of rainfall determining freshwater inflows and nutrient delivery, and from the bathymetric contours of the seafloor defining fish habitats to the coastal geomorphology dictating breeding grounds, these environmental forces are the silent yet powerful architects of aquatic life. It is precisely this undeniable and pervasive influence that makes the integrated study of meteorology and geography utterly essential for comprehending and managing our precious fishery resources.

Recognizing a critical **knowledge gap** in how these fundamental environmental sciences are systematically integrated into fisheries education, the **authors** have undertaken the ambitious yet crucial task of dissecting and synthesizing these pervasive influences. They have not only identified but brilliantly articulated how atmospheric and terrestrial processes are active protagonists, dictating the very existence, migration, and abundance of marine and freshwater organisms. Their meticulous work in this volume provides a much-needed bridge, seamlessly connecting these essential disciplines to the core narrative of fisheries management and research.

What truly elevates **Meteorology and Geography** is the **authors' remarkable ability to distill complex scientific principles into an eminently digestible format.** Far from being a mere compilation of facts, this textbook is a masterclass in pedagogical design. Every chapter reflects their dedication to **ease of learning,** presenting challenging concepts with such clarity and precision that they become accessible to students from diverse academic backgrounds. This is a direct result of their deep understanding of the subject matter coupled with an empathetic appreciation for the learning process.

The book's structure, which rigorously adheres to the **ICAR's Sixth Deans' Committee Report-based syllabus,** further underscores their commitment to academic excellence and practical utility.

This volume is poised to become an **invaluable resource** for the entire fisheries community. For **students**, it offers a comprehensive and engaging gateway to understanding the environmental underpinnings of their chosen field, equipping them with the analytical tools and foundational knowledge vital for navigating the complexities of modern fisheries. For **faculty**, it provides a meticulously crafted and authoritative teaching aid, empowering them to deliver impactful and relevant instruction that meets the highest educational standards. The authors have truly created a framework for both exploration and mastery.

I extend my sincerest commendation to the authors **Professors Surender Singh, V Uma Maheswara Rao and BV Ramana Rao** for their unparalleled dedication, scholarly rigor, and the immense effort they have invested in bringing this vital work to fruition. Their collaborative spirit and individual brilliance shine through on every page, creating a resource that is both academically robust and remarkably user-friendly.

I am immensely confident that **Meteorology and Geography** will not only enrich the academic journey of countless aspiring fisheries professionals but also ignite a deeper, more holistic understanding of the profound environmental forces that govern our planet's aquatic resources. It is a testament to the authors' unwavering commitment to education and sustainable resource management, and I wholeheartedly recommend it as an essential text for anyone passionate about the future of fisheries science.

(Mrutyunjaya Mohapatra)

Preface

The intricate interplay between our planet's atmosphere, its landforms, and the delicate balance of aquatic ecosystems is a subject of profound importance. In the realm of fisheries science, understanding this dance is not merely academic; it's fundamental to sustainable management and the continued health of our aquatic resources. With Meteorology and Geography, we aim to illuminate the indispensable role these twin disciplines play in unraveling the complexities of fish populations, their habitats, and the broader aquatic environment.

For too long, a discernible knowledgc gap has persisted in comprehensively integrating meteorological and geographical principles into fisheries curricula. While the impact of weather patterns on ocean currents, and the influence of coastal features on breeding grounds, are implicitly understood, a dedicated, integrated resource has been largely absent. This book was conceived to directly address that void, offering a holistic perspective on how atmospheric phenomena and terrestrial landscapes profoundly shape every aspect of fisheries science—from fish distribution and migration to productivity and vulnerability to environmental change.

We've meticulously structured this book to align perfectly with the syllabus prescribed by the ICAR's Sixth Deans' Committee Report. Our primary goal was to create a resource that not only meets rigorous academic standards but also fosters genuine ease of learning. We've strived to present complex scientific concepts in a clear, accessible, and engaging manner, utilizing straightforward language, illustrative examples, and practical applications. The content progresses logically, building foundational knowledge step-by-step to ensure a comprehensive and intuitive understanding for all readers.

Ultimately, Meteorology and Geography is designed to be an invaluable resource for both students and faculty. For students, it serves as a definitive guide, equipping them with the essential knowledge and analytical tools necessary to understand the environmental dimensions of fisheries. For our esteemed faculty colleagues, we hope it provides a well-organized, comprehensive, and adaptable teaching aid, facilitating robust and engaging discussions in the classroom.

We are deeply grateful to Dr Mrutyunjaya Mohapatra, Director General of Meteorology, India Meteorological Department, Ministry of Earth Sciences, Govt of India for graciously contributing such a insightful and compelling foreword to this volume. Their words truly set the tone and underscore the significance of the material within. We sincerely express our thanks to Shri Sumit Jain, Director-NIPA Genx Electronic Resources & Solutions Pvt for encouraging us to take up writing of the present book.

It is our sincere hope that this book will inspire a deeper appreciation for the intricate connections between our atmosphere, land, and aquatic life, fostering a new generation of fisheries professionals who are well-equipped to navigate the challenges and opportunities of a changing world. We believe this textbook will be an indispensable companion in the pursuit of knowledge and the advancement of sustainable fishery practices.

Surender Singh
V Uma Maheswara Rao
BV Ramana Rao

Acknowledgement

It gives us immense pleasure to present this book titled 'Meteorology and Geography', intended to support students in understanding the vital link between atmospheric sciences, geographical patterns, and fisheries.

We express my deep gratitude to the academic and scientific community whose foundational work and literature guided the development of this text. We are especially thankful to our colleagues and associates in the fields of Agrometeorology, Geography, and Fisheries Science for their insightful discussions, critical reviews, and constant encouragement throughout this endeavor.

Our sincere thanks go to the faculty members and students of Fisheries Science programs across universities, whose feedback and classroom experiences have helped shape the content and pedagogy of this book. Their curiosity and commitment to learning remain a source of inspiration. We are grateful to the publishing team for their professionalism and timely cooperation.

Lastly, We extend heartfelt thanks to our family for their unwavering support and patience during the long hours of writing and editing.

Any errors or omissions are solely ours and we welcome constructive feedback from readers for future improvements.

Surender Singh
V Uma Maheswara Rao
BV Ramana Rao

Contents

List of Figures

1

Unveiling the Interplay of Atmosphere, Land, and Water for Fisheries

This book, "Meteorology and Geography" is designed to equip aspiring fisheries professionals with a foundational understanding of the atmospheric and terrestrial processes that profoundly influence aquatic ecosystems and, consequently, the distribution, abundance, and health of fish and other commercially important species. For too long, the study of fisheries has sometimes been perceived as largely focused on the biology and ecology within the water itself. While these aspects are undeniably vital, the reality is that aquatic environments are not isolated entities. They are inextricably linked to the atmosphere above and the land surrounding them. The weather patterns we experience daily – from sunshine and rainfall to storms and temperature fluctuations – directly impact the physical and chemical characteristics of our oceans, lakes, and rivers. Similarly, the geographical landscape, including topography, soil composition, and coastal features, plays a critical role in shaping these aquatic habitats.

Consider for a moment the seemingly simple act of rain. Rainfall over land influences river discharge, bringing nutrients and sediments into freshwater and coastal systems. Changes in salinity, turbidity, and nutrient levels can dramatically alter the suitability of these waters for different fish species. Strong winds can drive ocean currents, upwelling events that bring nutrient-rich deep waters to the surface, and the dispersal of fish larvae. Extreme weather events like cyclones and floods can cause devastating impacts on both wild and aquaculture-based fisheries.

Furthermore, the broader geographical context dictates the regional climate, influencing long-term patterns of water temperature, precipitation, and wind. These climatic conditions, in turn, determine the types of aquatic ecosystems that can thrive in a particular area and the species that inhabit them.5 Understanding these geographical controls is essential for predicting the potential and limitations of fisheries in different regions.

The book will guide you through the fundamental principles of meteorology and physical geography, specifically highlighting their relevance to fisheries science and management. We will explore:

- **The Earth's Atmosphere:** Its composition, structure, and the processes that drive weather phenomena such as temperature variations, wind patterns, precipitation, and storms. We will delve into how these atmospheric conditions directly and indirectly affect aquatic environment).
- **Physical Geography:** The Earth's landforms, processes shaping them (like erosion and deposition), and the distribution of water bodies. We will examine how these geographical features influence coastal environments, river systems, and lake basins, ultimately impacting fish habitats.
- **Oceanography (with a meteorological and geographical lens):** While not solely a meteorology and geography text, we will explore key oceanographic concepts driven by atmospheric forces (like currents and upwelling) and influenced by coastal geography.
- **The Interconnections:** We will emphasize the crucial linkages between atmospheric and terrestrial processes and their cascading effects on aquatic ecosystems and fisheries. This includes topics like the impact of land use on water quality and the influence of climate change on fisheries resources.

By the end of this book, you will gain a holistic perspective, recognizing that the success and sustainability of fisheries are not solely dependent on what happens within the water but are significantly influenced by the dynamic interplay of the atmosphere and the land. This knowledge will be invaluable as you pursue careers in fisheries management, aquaculture, research, and conservation.

We encourage you to approach this material with curiosity and a desire to connect the seemingly disparate fields of meteorology, geography, and fisheries. The insights you gain here will provide a crucial foundation for understanding the complexities of aquatic ecosystems and for making informed decisions towards the responsible utilization and conservation of our precious aquatic resources.

Let us embark on this journey to unravel the intricate connections between the sky, the land, and the waters that sustain life and livelihoods.

1.1 Meteorology

Meteorology, in the context of fisheries and aquaculture, is the scientific study of the Earth's atmosphere, specifically examining how weather and climate influence aquatic environments and the organisms living within them. This includes understanding:

- **Direct impacts:** How temperature fluctuations, wind patterns, precipitation (rainfall and snowfall), solar radiation, and extreme weather events (storms, cyclones, heatwaves) directly affect water temperature, salinity, oxygen levels, water currents, and overall habitat suitability for fish and other aquaculture species.
- **Indirect impacts:** How atmospheric conditions influence terrestrial processes that subsequently affect aquatic ecosystems. For example, rainfall patterns dictate river discharge, which carries nutrients and sediments into water bodies, impacting water quality and productivity. Wind can drive mixing in lakes and oceans, influencing nutrient distribution and plankton blooms, which form the base of many aquatic food webs.
- **Long-term climate change:** How shifts in global and regional climate patterns, such as rising water temperatures, changes in precipitation regimes, and increased frequency of extreme weather, pose significant challenges and opportunities for both wild fisheries and aquaculture operations.

Understanding meteorological principles is crucial for:

- **Predicting and mitigating risks:** Forewarning of storms, extreme temperatures, or changes in water quality that could harm aquaculture stocks or disrupt fishing activities.
- **Optimizing aquaculture practices:** Selecting suitable locations, managing water quality parameters (temperature, dissolved oxygen), and timing feeding schedules based on environmental conditions.
- **Managing wild fisheries:** Understanding how climate variability and long-term climate change affect fish distribution, migration patterns, and reproductive success.

In essence, meteorology provides a vital lens through which we can understand the dynamic interactions between the atmosphere and the aquatic realm, which is fundamental to the sustainable management and development of fisheries and aquaculture.

1.2 Earth's Atmosphere: Definition, Detailed Vertical Structure, and Composition

What is the Atmosphere?

The Earth's atmosphere is the blanket of gases surrounding our planet, held in place by gravity. It's a complex, life-sustaining system that provides breathable air, shields us from harmful solar and cosmic radiation, and regulates Earth's temperature, making life as we know it possible (Fig.1.1).

Fig. 1.1: Atmosphere - A vast and dynamic fluid that surrounds planet earth

1.2.1 Composition of Weather and Climate

The composition of Earth's atmosphere is a delicate balance of several gases, with regional and temporal variations (Fig. 1.2). The major components of dry air are:

- **Nitrogen (N_2):** Approximately 78%. Nitrogen is relatively inert and plays a crucial role in diluting oxygen, preventing rapid combustion.
- **Oxygen (O_2):** Approximately 21%. Oxygen is vital for respiration in most living organisms and is essential for combustion. It's primarily produced by photosynthesis.
- **Argon (Ar):** Approximately 0.9%. Argon is an inert noble gas.

In addition to these major components, the atmosphere contains several variable components that play critical roles:

- **Water Vapor (H_2O):** Its concentration varies widely (0-4%) depending on location and time. Water vapor is crucial for humidity, cloud formation, and precipitation.
- **Carbon Dioxide (CO_2):** A trace gas (around 0.04% or 400 parts per million), but its concentration is increasing due to human activities. CO_2

is a greenhouse gas, playing a vital role in regulating Earth's temperature and is crucial for photosynthesis.

- **Methane (CH_4):** Another potent greenhouse gas, present in small amounts. Methane is emitted from natural sources (wetlands, decomposition) and human activities (agriculture, fossil fuel production).
- **Ozone (O_3):** Concentrated in the stratosphere, the ozone layer absorbs most of the Sun's harmful ultraviolet (UV) radiation.
- **Aerosols:** Tiny solid and liquid particles (dust, sea salt, volcanic ash, pollution) that influence cloud formation and affect the amount of sunlight that reaches the surface.

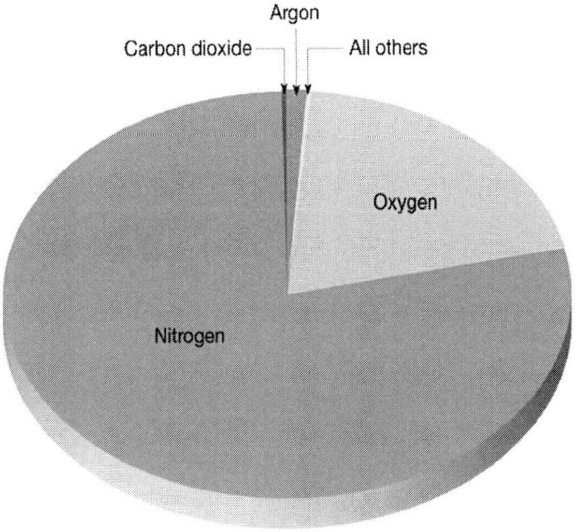

Fig. 1.2: Composition of Earth's atmosphere

1.2.2 Vertical Structure of the Atmosphere

The atmosphere is divided into distinct layers based on how temperature changes with altitude (Fig.1.3).

- **Troposphere**
  - **Altitude:** Extends from the Earth's surface up to an average of 12 km (about 7.5 miles). This height varies with latitude, being lower at the poles (around 8 km) and higher at the equator (around 18 km).
  - **Temperature:** Generally decreases with altitude at an average rate of 6.5 °C per kilometer. The warmest part is near the surface, heated by the Earth itself.

- **Key Features:** This is the densest layer, containing about 75% of the atmosphere's mass and almost all of its water vapor. All weather phenomena, such as clouds, rain, and storms, occur in the troposphere. Airplanes typically fly in the upper troposphere to avoid turbulence. The top of the troposphere is called the **tropopause**.

• **Stratosphere**

- **Altitude:** Extends from the tropopause up to about 50 km (about 31 miles).
- **Temperature:** Increases with altitude. This temperature inversion is due to the absorption of the Sun's ultraviolet (UV) radiation by the **ozone layer**, which is concentrated in this regio on Earth. Commercial jets often cruise in the lower stratosphere for a smoother ride. The top of the stratosphere is called the **stratopause**.

• **Mesosphere**

- **Altitude:** Extends from the stratopause up to about 85 km (about 53 miles).
- **Temperature:** Decreases with altitude, making it the coldest layer of the atmosphere. Temperatures can drop to around −90 °C near the top (the mesopause).
- **Key Features:** The mesosphere protects the Earth from space debris, as most meteors burn up in this layer due to friction with the atmospheric gases. Noctilucent clouds, the highest clouds in Earth's atmosphere, sometimes form near the mesopause.

• **Thermosphere**

- **Altitude:** Extends from the mesopause upwards to between 500 and 1,000 km (about 311 to 621 miles).
- **Temperature:** Increases dramatically with altitude due to the absorption of high-energy solar radiation (X-rays and UV). Temperatures can reach hundreds or even thousands of degrees Celsius, although it would not feel hot due to the extremely low air density.
- **Key Features:** The **ionosphere**, a region where gases are ionized by solar radiation, is located within the thermosphere (and extends into parts of the mesosphere and exosphere). The ionosphere reflects radio waves and is where auroras (the Northern and Southern Lights) occur. The International Space Station orbits in the thermosphere. The upper boundary is called the **thermopause**.

- **Exosphere**
  - **Altitude:** The outermost layer, starting from the thermopause and extending outwards into space. There is no sharp upper boundary.
  - **Temperature:** Very high in the lower exosphere, but the concept of temperature is less meaningful due to the extremely low density. Atoms and molecules are so far apart they rarely collide.
  - **Key Features:** The exosphere is composed of very sparse atoms and molecules, mainly hydrogen and helium, which can escape into space. Many satellites orbit in this layer. It represents the transition zone between Earth's atmosphere and outer space.

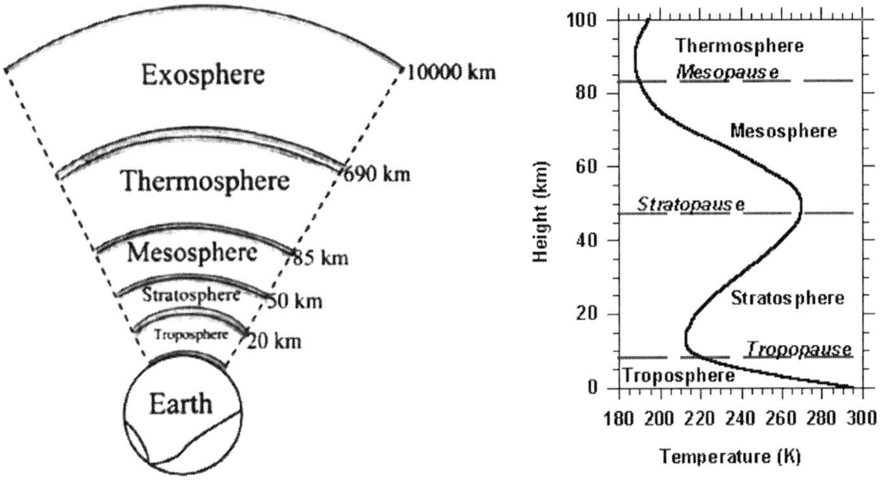

Fig. 1.3: Vertical structure of the atmosphere

1.3 Weather and Climate

Let us understand about **weather and climate**, their elements, and the factors that control climate (Fig. 1.4), keeping the relevance to fisheries and aquaculture in mind.

It's crucial to distinguish between weather and climate:

- **Weather** refers to the short-term (minutes to weeks) conditions of the atmosphere at a specific place and time. It's what you experience day-to-day – is it sunny, rainy, windy, hot, or cold *today*?
- **Climate**, on the other hand, describes the long-term (typically 30 years or more) average of weather patterns in a particular region. It's the general atmospheric conditions you expect – for example, a tropical climate is generally hot and humid year-round.

1.3.1 Relevance to Fisheries and Aquaculture

- **Weather** directly impacts daily operations. A sudden storm can disrupt fishing activities, damage aquaculture infrastructure (like cages or ponds), and rapidly change water conditions (temperature, turbidity, salinity in coastal areas). Daily temperature and sunlight affect metabolic rates and feeding in aquaculture.
- **Climate** dictates the overall suitability of a region for certain aquatic species, both wild and farmed. Long-term changes in climate, like warming trends, can shift species distributions, alter breeding cycles, and increase the frequency of extreme events that impact aquatic ecosystems and aquaculture.

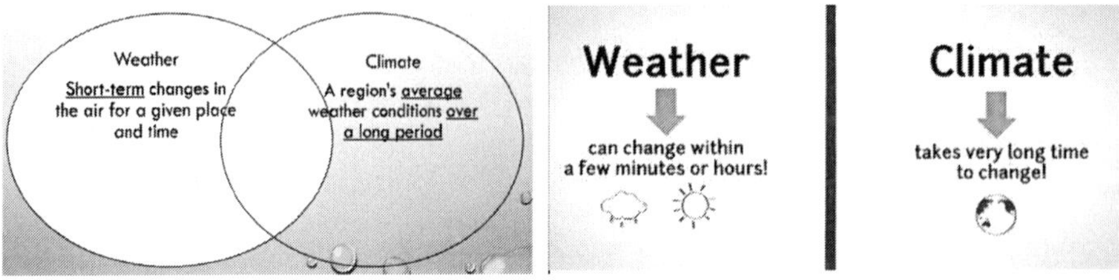

Fig. 1.4: Weather and climate

1.3.2 Weather Elements

These are the measurable atmospheric properties that describe the current state of the weather. Key weather elements include:

- **Temperature:** The degree of hotness or coldness of the air. Directly affects water temperature, influencing fish physiology and growth.
- **Atmospheric Pressure:** The weight of the air above a certain point. Changes in pressure are often associated with changes in weather systems (e.g., low pressure can bring storms). Can indirectly affect water levels in some systems.
- **Wind (Speed and Direction):** The movement of air. Affects wave action, mixing of water layers (important for oxygen and nutrient distribution), and can influence the dispersal of fish larvae or pollutants. Strong winds can also be a hazard to aquaculture structures.
- **Humidity:** The amount of water vapor in the air. High humidity can lead to fog, which can reduce sunlight penetration into water, affecting primary production.
- **Precipitation (Type and Amount):** Any form of water that falls from clouds to the Earth's surface (rain, snow, sleet, hail). Rainfall affects river

discharge, salinity in estuaries, and can carry nutrients and pollutants into water bodies. Snowmelt is a significant source of freshwater in many regions.

- **Cloud Cover (Type and Amount):** Clouds affect the amount of solar radiation reaching the water surface, influencing water temperature and light availability for photosynthesis.
- **Sunshine Duration:** The length of time the Earth's surface is directly exposed to solar radiation. Impacts water temperature and primary productivity.
- **Visibility:** The distance at which an object can be clearly seen. Important for safe navigation for fishing vessels and can be reduced by fog (related to humidity and temperature).

1.3.3 Factors Controlling Climate

These are the relatively constant influences that determine the long-term average weather patterns of a region. The main factors include:

- **Latitude:** Distance from the equator. Affects the angle of incoming solar radiation and thus the amount of solar energy received. Lower latitudes (closer to the equator) generally receive more direct sunlight and have warmer climates, which influences the types of aquatic species that thrive there.
- **Altitude (Elevation):** Height above sea level. Temperature generally decreases with increasing altitude. This affects the temperature of freshwater systems at different elevations, influencing the distribution of cold-water versus warm-water fish species.
- **Distance from the Sea (Continentality):** Large bodies of water moderate temperatures. Coastal areas tend to have smaller temperature ranges (cooler summers, warmer winters) compared to inland areas at the same latitude. This maritime influence affects the temperature regimes of coastal and estuarine environments, which are critical for many fisheries.
- **Ocean Currents:** The movement of large volumes of ocean water. Warm currents can warm coastal regions, while cold currents can cool them. These currents also influence nutrient distribution and the distribution of marine species. Upwelling, often driven by currents and wind patterns, brings nutrient-rich cold water to the surface, supporting productive fisheries.

- **Prevailing Winds:** The dominant direction from which the wind blows in a region. Winds can transport air masses with different temperature and humidity characteristics, influencing regional climate and also affecting surface currents in water bodies.
- **Topography (Relief):** The shape of the land, including mountains. Mountains can create rain shadows (dry areas on the leeward side) and influence local temperature and precipitation patterns, affecting freshwater habitats and runoff into coastal areas.

1.3.4 In the context of fisheries and aquaculture, understanding these factors is vital for

- **Site selection for aquaculture:** Choosing locations with suitable temperature regimes, water availability, and protection from extreme weather influenced by these factors.
- **Predicting the impact of climate change:** Assessing how shifts in these controlling factors (e.g., rising global temperatures, changes in ocean currents) will affect aquatic ecosystems and fisheries resources.
- **Managing wild fisheries:** Understanding how regional climate influences the distribution, abundance, and productivity of fish stocks.

1.4 Heat Energy in the Atmosphere

Let's delve into the **heat energy of the atmosphere** in detail, with a focus on its relevance to aquatic environments and, by extension, fisheries and aquaculture.

The atmosphere's heat energy is the foundation of our planet's climate and weather systems. It drives winds, precipitation, and temperature patterns, all of which have profound effects on aquatic ecosystems.

1.4.1 Sources of the Atmosphere's Heat and Heating of the Atmosphere

The primary source of heat for the Earth's atmosphere is the **Sun**.However, the atmosphere isn't heated directly by the incoming solar radiation as much as you might think. Here's a breakdown:

- **Solar Radiation (Insolation):** The Sun emits a vast amount of electromagnetic radiation, a portion of which reaches the Earth. This incoming solar radiation, or insolation, is shortwave radiation (including visible light and ultraviolet).
    - About 30% of this incoming radiation is reflected back to space by clouds, the atmosphere itself, and the Earth's surface (especially ice and snow - this reflectivity is called **albedo**).

- Roughly 20% is absorbed directly by the atmosphere through gases like ozone (absorbing UV), water vapor, and dust.
- The remaining 50 % reaches the Earth's surface (land and oceans).

- **Terrestrial Radiation:** The Earth's surface, having absorbed solar radiation, warms up and then re-radiates energy back into the atmosphere. However, because the Earth is much cooler than the Sun, it emits this energy as **longwave infrared radiation** (heat). This is the *primary* way the lower atmosphere is heated.
  - Certain gases in the atmosphere, known as **greenhouse gases** (water vapor, carbon dioxide, methane, etc.), are very effective at absorbing this outgoing longwave radiation.
  - Once absorbed, these gases re-emit the energy in all directions, some of which goes back towards the Earth's surface, contributing to the warming of the lower atmosphere – this is the **greenhouse effect**, a natural process crucial for maintaining a habitable temperature on Earth.

1.4.2 Transfer of Heat in the Atmosphere

Once heat energy enters the atmosphere, it is distributed through several processes:

- **Radiation:** As discussed, this is the transfer of energy through electromagnetic waves. The Sun's energy reaching Earth and the Earth's energy radiating back are examples. The atmosphere also radiates energy out to space.
- **Conduction:** This is the transfer of heat through direct contact. The air in direct contact with the warm Earth's surface gets heated through conduction. However, air is a poor conductor, so this process is most effective in the layers of air closest to the surface.
- **Convection:** This involves the transfer of heat through the movement of fluids (like air). Warm air is less dense and rises, carrying heat upwards. Cooler, denser air sinks to replace it, creating vertical air currents. This is a very important mechanism for transferring heat throughout the troposphere (the lowest layer of the atmosphere). Think of thunderstorms, where warm, moist air rises rapidly.
- **Advection:** This is the horizontal transfer of heat by the movement of air masses. Winds carrying warmer or colder air from one region to another are examples of advection.

1.4.3 Relevance to Fisheries and Aquaculture

Understanding the heat energy of the atmosphere and its distribution is crucial for fisheries students because it directly and indirectly influences aquatic environments:

- **Water Temperature:** Atmospheric temperature strongly influences surface water temperatures in lakes, rivers, and oceans. Warmer air generally leads to warmer water, affecting the physiology, growth rates, reproduction, and distribution of aquatic species. In aquaculture, temperature control is vital for optimal growth and health of farmed organisms.
- **Dissolved Oxygen:** Water temperature affects the solubility of gases, including oxygen. Warmer water holds less dissolved oxygen, which can stress or even kill aquatic life. Atmospheric temperature, by influencing water temperature, indirectly impacts dissolved oxygen levels.
- **Stratification:** In lakes and sometimes coastal areas, temperature differences can lead to stratification (layering) of the water column. This stratification, driven by surface heating or cooling influenced by atmospheric conditions, can affect nutrient distribution and oxygen availability at different depths, impacting where fish can live.
- **Ice Formation and Thawing:** In colder regions, atmospheric temperatures determine when and how extensively ice forms on water bodies. Ice cover affects light penetration (important for primary producers), gas exchange, and habitat availability for fish. Thawing patterns, also driven by atmospheric warming, influence spring runoff and nutrient inputs.
- **Extreme Weather Events:** Storms and heatwaves, driven by atmospheric energy imbalances, can have devastating impacts on fisheries and aquaculture. Storm surges can flood coastal aquaculture facilities, while prolonged high atmospheric temperatures can lead to dangerously high water temperatures.
- **Climate Change:** Long-term changes in the atmosphere's heat balance, primarily due to the enhanced greenhouse effect, are leading to significant changes in water temperatures, ocean acidification, sea-level rise, and altered weather patterns, all of which have profound implications for the future of fisheries and aquaculture.

In summary, the heat energy of the atmosphere is a fundamental driver of environmental conditions that directly shape the habitats and well-being of aquatic organisms. As future fisheries professionals, understanding these atmospheric processes will be essential for sustainable management and responsible aquaculture practices.

1.4.4 Relevance to Aquatic Environments

In summary, the heating of the atmosphere is a multi-stage process starting with solar radiation, followed by the Earth's surface warming and emitting longwave radiation, which is then absorbed by greenhouse gases.This heat is then distributed through conduction, convection, and advection, all of which have direct and indirect impacts on the temperature and other physical characteristics of aquatic environments crucial for fisheries and aquaculture.

1. Unveiling the Interplay of Atmosphere, Land, and Water for Fisheries

1. **Which atmospheric element most directly affects dissolved oxygen in water?**
   A. Temperature
   B. Humidity
   C. Wind
   D. Pressure
2. **What is the primary source of energy that drives atmospheric and aquatic processes?**
   A. Moonlight
   B. Geothermal energy
   C. Solar radiation
   D. Wind power
3. **Low-pressure systems typically lead to:**
   A. Clear skies and stable conditions
   B. Increased rainfall and wind
   C. Freezing temperatures
   D. High salinity in water
4. **Wind patterns affect water bodies primarily through:**
   A. Evaporation
   B. Salinity control
   C. Mixing and aeration
   D. Sediment compaction
5. **Which gas increases in solubility in water as temperature decreases?**
   A. Nitrogen
   B. Oxygen
   C. Carbon dioxide
   D. Methane
6. **Stable weather conditions with little atmospheric disturbance benefit fisheries by:**
   A. Enhancing nutrient runoff
   B. Promoting water column stratification
   C. Increasing turbidity
   D. Reducing photosynthesis
7. **Which phenomenon often results in fish kill due to sudden change in water temperature?**
   A. Cold shock
   B. Algal bloom
   C. Acid rain
   D. Solar eclipse

8. Atmospheric humidity plays a role in:

A. Oxygen uptake by fish
B. Water temperature regulation
C. Precipitation and runoff
D. Increasing salinity

9. Which atmospheric layer is most relevant to fishery meteorology?

A. Mesosphere
B. Stratosphere
C. Troposphere
D. Thermosphere

10. Which is an extreme weather event influenced by atmospheric instability?

A. Harmattan
B. Tsunami
C. Cyclone
D. Earthquake

11. Which land feature significantly impacts watershed runoff into fish habitats?

A. Mountains
B. Forests
C. Soil type
D. All of the above

12. Soil erosion in catchment areas leads to:

A. Increased oxygen in water
B. Sediment loading in aquatic systems
C. Decreased nutrient inflow
D. Higher water transparency

13. Agricultural land use can increase nutrient input into water bodies, resulting in:

A. Oligotrophication
B. Eutrophication
C. Salinization
D. Acidification

14. Which land-based activity most commonly pollutes freshwater fisheries?

A. Mining
B. Urbanization
C. Deforestation
D. All of these

15. What is the role of riparian vegetation in fisheries?

A. Adds salinity
B. Stabilizes banks and filters runoff
C. Increases evaporation
D. Prevents fish migration

16. Deforestation upstream can result in:

A. Improved fish habitat
B. Reduced sediment load
C. Flash floods
D. Enhanced oxygen content

17. Pesticide runoff from land mainly affects fish by:

A. Boosting growth
B. Causing mutation
C. Bioaccumulation and toxicity
D. Increasing fertility

18. Landform slope influences:

A. Oxygen exchange
B. Wind patterns
C. Rate of surface runoff
D. Water temperature

19. Wetlands between land and water serve as:

A. Barriers to nutrient flow
B. Filters and breeding grounds
C. Sources of salt intrusion
D. None of the above

20. Which soil characteristic impacts groundwater recharge for fish ponds?

A. Color
B. Structure
C. Permeability
D. Texture only

21. Which water parameter is most vital for fish respiration?

A. Salinity
B. Dissolved Oxygen
C. Ammonia
D. Carbonate

22. Increased water turbidity can result in:

A. Enhanced photosynthesis
B. Light penetration reduction
C. More predators
D. Temperature stabilization

23. Thermal stratification in lakes is influenced by:

A. Wind speed only
B. Sunlight intensity
C. Depth and temperature
D. Algal presence

24. A major consequence of eutrophication is:

A. Clearer water
B. Higher fish yield
C. Algal blooms and hypoxia
D. Increase in crustaceans

25. pH of water ideal for most freshwater fish is:

A. 3.5–4.5
B. 5.0–6.0
C. 6.5–8.5
D. 9.0–11.0

26. Salinity fluctuations mainly impact:

A. Warm-water species
B. Deep-sea fish
C. Estuarine species
D. Herbivorous fish

27. Which nutrient is most often responsible for algal blooms?

A. Nitrogen
B. Phosphorus
C. Potassium
D. Sulphur

28. Which waterborne element is toxic in high concentration to fish?

A. Calcium
B. Magnesium
C. Ammonia
D. Chloride

29. Biochemical oxygen demand (BOD) is an indicator of:

A. Salinity
B. Organic pollution
C. Turbidity
D. Temperature

30. Which type of aquatic habitat is most sensitive to pollution?

A. Brackish
B. Marine
C. Estuarine
D. Freshwater

31. Seasonal rainfall influences fisheries by:

A. Reducing primary production
B. Regulating water availability and breeding
C. Stabilizing temperature
D. Diluting salinity only

32. El Niño can impact marine fisheries by:

A. Cooling coastal waters
B. Increasing oxygen in deep waters
C. Altering upwelling and fish stocks
D. Improving coral reef health

33. Which climatic parameter best predicts fish migration patterns?

A. Rainfall
B. Solar radiation
C. Wind direction
D. Water temperature

34. Extreme drought conditions affect inland fisheries by:

A. Enhancing productivity
B. Lowering water levels and increasing stress
C. Reducing oxygen stress
D. Stimulating fish spawning

35. Cyclones are particularly hazardous for:

A. Marine ecosystems only
B. Fish hatcheries near coastlines
C. Deep-sea aquaculture
D. Cold-water fish

36. The monsoon season in India is critical for fish farming due to:

A. Increased salinity
B. Enhanced breeding and pond filling
C. Acid rain risk
D. Higher pest infestation

37. Climate change impacts fisheries through:

A. Sea level rise
B. Ocean acidification
C. Changing temperature regimes
D. All of the above

38. Which species is most vulnerable to temperature fluctuations?

A. Carp
B. Tilapia
C. Trout
D. Catfish

39. Cloudy days reduce fish activity primarily due to:

A. Decrease in water pressure
B. Lower light for primary producers
C. Increase in predators
D. Sudden rise in pH

40. Anomaly in fish migration can be attributed to:

A. Stable salinity
B. Consistent wind
C. Temperature shocks
D. Higher productivity

41. Which model integrates atmospheric and hydrological data for fisheries planning?

A. GIS | B. DSSAT
C. SWAT | D. CMIP

42. Watershed management helps fisheries by:

A. Reducing water storage | B. Increasing runoff
C. Controlling erosion and pollutants | D. Improving wind speed

43. Rainwater harvesting near ponds helps in:

A. Increasing temperature | B. Enhancing oxygen diffusion
C. Replenishing water supply | D. Lowering pH

44. Which policy measure supports climate-resilient fisheries?

A. Water export | B. Use of pesticides
C. Early warning systems | D. Fishing bans in summer

45. The combination of meteorological and hydrological inputs is called:

A. Oceanology | B. Hydroclimatology
C. Aquaculture management | D. Ecometeorology

46. Best way to control pollution in fisheries sector is:

A. Increase fish feed | B. Treat wastewater before discharge
C. Add chemicals to neutralize | D. Use deeper ponds

47. GIS in fisheries is mainly used for:

A. Depth estimation | B. Mapping habitats and water bodies
C. Calculating BOD | D. Predicting rainfall

48. Fish stress can be reduced by:

A. Maintaining water quality parameters | B. Increasing light exposure
C. Using herbicides | D. Harvesting early

49. Which indicator is used for aquatic ecosystem health?

A. GDP | B. Chlorophyll concentration
C. Wind speed | D. Soil color

50. Interdisciplinary study of atmosphere, land, and water in fisheries is essential for:

A. Technology development only | B. Better taxation
C. Holistic ecosystem management | D. Weather prediction

Answer Key (s)

1	a	2	c	3	b	4	c	5	b	6	b	7	a	8	c	9	c	10	c
11	d	12	b	13	b	14	d	15	b	16	c	17	c	18	c	19	b	20	c
21	b	22	b	23	c	24	c	25	c	26	c	27	b	28	c	29	b	30	d
31	b	32	c	33	d	34	b	35	b	36	b	37	d	38	c	39	b	40	c
41	c	42	c	43	c	44	c	45	c	46	b	47	b	48	a	49	b	50	c

2

Insolation: The Sun's Energy and Aquatic Life

Insolation is a term that might sound abstract but is fundamentally the driving force behind nearly all physical processes on Earth, including those within our precious aquatic environments. In this chapter, we will unravel the journey of solar energy from the vast expanse of space to the surfaces of our oceans, lakes, and rivers, and understand its profound influence on the life within.

For you, as future professionals in fisheries and aquaculture, grasping the concept of insolation is paramount. The amount of solar radiation reaching a water body dictates its surface temperature, influences the rate of photosynthesis by aquatic plants and algae (the base of many aquatic food webs), affects water stratification, and even plays a role in phenomena like ice formation and melting.

Think about it: the warmth of a lake on a sunny day, the intensity of light penetrating the surface waters, the seasonal changes in algal blooms – all are directly linked to the amount and intensity of solar radiation received. Understanding insolation allows us to begin to explain why some aquatic environments are more productive than others, why water temperatures vary geographically and seasonally, and how changes in solar energy input (whether natural or human-induced) can ripple through entire aquatic ecosystems.

In this chapter, we will delve into

- **What insolation is:** Defining the term and understanding its measurement.
- **Factors affecting insolation:** Exploring how latitude, time of year, time of day, and atmospheric conditions (like clouds) influence the amount of solar radiation reaching the Earth's surface and water bodies.
- **The fate of insolation upon reaching water:** Examining how solar energy is absorbed, reflected, and transmitted within aquatic environments.

- **The crucial links to aquatic life:** Discussing how insolation directly and indirectly impacts water temperature, primary productivity, dissolved oxygen, and the overall health and functioning of aquatic ecosystems relevant to fisheries and aquaculture.

By the end of this chapter, you will appreciate that the seemingly distant energy from the Sun is a primary driver of the conditions that govern the lives of the fish and other aquatic organisms you will be studying and managing. Let's embark on understanding this fundamental energy source that shapes the aquatic world.

2.1 Insolation

Insolation is the amount of solar radiation (energy from the sun) that reaches a specific area on the Earth's surface over a given period (Fig.2.1). The term itself is a combination of "incoming solar radiation."

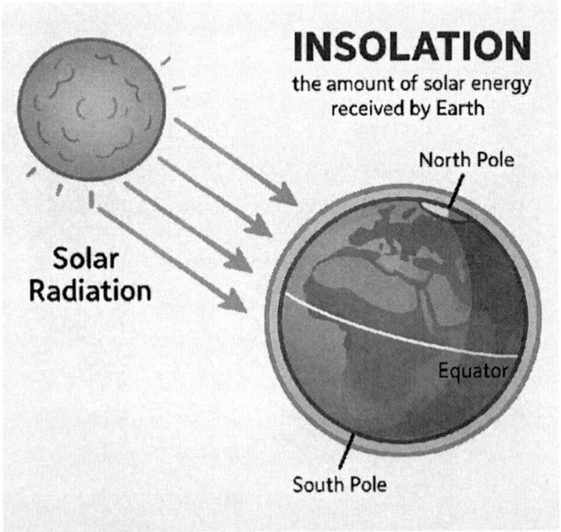

Fig. 2.1: Insolation/solar radiation

Think of it as the sunlight arriving at a particular spot on our planet. This energy is crucial because it drives many of Earth's systems, including the heating of the atmosphere and oceans, which are fundamental to aquatic life and thus, fisheries and aquaculture.

The amount of insolation received at any location varies based on several factors, including:

- **Latitude:** Regions closer to the equator receive more direct sunlight and thus higher insolation compared to areas at higher latitudes (closer to the poles).

- **Time of year:** Due to the Earth's tilt, different parts of the Earth receive varying amounts of direct sunlight throughout the year, causing seasons.
- **Time of day:** Insolation is highest around noon when the sun is highest in the sky.
- **Atmospheric conditions:** Clouds, dust, and other particles in the atmosphere can reflect, scatter, and absorb solar radiation, reducing the amount that reaches the surface.

For fisheries students, understanding insolation is key because it directly influences:

- **Water temperature:** More insolation generally leads to warmer surface waters.
- **Primary productivity:** Sunlight is essential for photosynthesis by phytoplankton and aquatic plants, which form the base of the aquatic food web.
- **Water stratification:** Differences in surface heating due to insolation can lead to the formation of layers in water bodies with different temperatures and densities.

2.1.1 Disposal of insolation

Okay, let's discuss the "disposal" or, more accurately, what happens to insolation when it reaches the Earth's surface, particularly focusing on water bodies relevant to fisheries and aquaculture. When solar radiation encounters a body of water, several things can happen:

- **Reflection:** Some of the insolation bounces off the water's surface. The amount reflected depends on factors like the angle of the sun (more is reflected at lower angles), the wavelength of light, and the smoothness of the water surface. Typically, for a high sun angle, only about 5-10% of the incoming solar radiation is reflected from the water surface.
- **Absorption:** A significant portion of the insolation penetrates the water surface and is absorbed. Water molecules absorb different wavelengths of light at different rates.
    - **Infrared radiation** is absorbed very quickly in the top layer of water, contributing significantly to surface heating. About 90% of infrared radiation is absorbed within the first meter.
    - **Visible light** penetrates deeper. Longer wavelengths (red, orange) are absorbed more readily than shorter wavelengths (blue, green), which can penetrate to greater depths. This differential absorption affects the light available for photosynthesis at different depths.

- **Ultraviolet (UV) radiation** is also absorbed, with shorter wavelengths (UV-C, UV-B) being absorbed more readily in the upper layers than UV-A, which can penetrate deeper. The absorbed solar radiation is converted into heat, increasing the water's temperature.
- **Transmission:** Some solar radiation passes through the water. The amount of transmission depends on the water's clarity (turbidity) and depth. In clear water, some light can penetrate to considerable depths. This transmitted light is crucial for photosynthesis by submerged aquatic vegetation and phytoplankton.

2.1.2 Relevance to Fisheries and Aquaculture

- **Water Temperature:** The absorbed component of insolation is the primary driver of water temperature. Surface waters heat up more quickly due to the high absorption of infrared radiation. This temperature directly affects the metabolic rates, growth, and reproductive cycles of aquatic organisms.
- **Light Penetration:** The amount and type of light that penetrates the water column (the transmitted portion of insolation, minus what's absorbed along the way) is vital for primary producers (algae and aquatic plants). Photosynthesis, the foundation of many aquatic food webs, is dependent on sufficient light.
- **Thermal Stratification:** Differential heating due to the absorption of insolation with depth can lead to the formation of distinct temperature layers in lakes and oceans, affecting oxygen levels and habitat availability for different species.
- **UV Radiation Effects:** The absorption of UV radiation in the upper layers can impact plankton and fish larvae, potentially causing harm depending on the intensity and duration of exposure.

In essence, understanding how insolation is "disposed of" when it reaches water is fundamental to understanding the physical and biological conditions within aquatic ecosystems that directly influence fisheries and aquaculture.

2.2 Irregular Heating of the Aatmosphere

Let's discuss **irregular heating of the atmosphere**: what it is and how it happens, with a focus on its implications for aquatic environments (Fig.2.2).

Fig. 2.2: Solar radiation and fisheries

What is Irregular Heating of the Atmosphere?

The Earth's atmosphere is not heated uniformly. Some regions receive more solar energy and retain more heat than others. This **irregular heating** refers to these spatial differences in air temperature across the globe and even at local levels.

• How Does Irregular Heating Happen?

Several factors contribute to this uneven distribution of heat:

• Spherical Shape of the Earth and Earth's Tilt

- The Earth is a sphere, so different latitudes receive solar radiation at varying angles. The equator receives more direct, concentrated sunlight, leading to greater heating. Towards the poles, the sunlight strikes at a lower angle, spreading the energy over a larger area, resulting in less heating.
- The Earth's axial tilt (about 23.5 degrees) causes seasons. During different times of the year, different hemispheres are tilted more directly towards the sun, leading to greater insolation and thus more heating in those regions.

- **Differential Absorption and Reflection (Albedo)**
  - The Earth's surface is not uniform. Land, water, ice, forests, and deserts have different abilities to absorb and reflect solar radiation. This reflectivity is called **albedo**.
  - Darker surfaces (like oceans and forests) absorb more solar radiation and heat up more than lighter surfaces (like ice and deserts), which reflect more energy back into space.6 This difference in absorption leads to temperature variations in the overlying atmosphere.
- **Uneven Distribution of Land and Water**
  - Land heats up and cools down faster than water due to water's higher specific heat capacity and its ability to mix. This leads to temperature differences between continental and oceanic regions at the same latitude.
- **Atmospheric Composition**
  - The distribution of greenhouse gases, clouds, and aerosols (tiny particles) in the atmosphere is not uniform. These components absorb and reflect solar and terrestrial radiation to varying degrees, contributing to regional differences in heating. For example, areas with more cloud cover might receive less solar radiation at the surface but also trap more outgoing longwave radiation at night.

2.2.1 Implications for Fisheries and Aquaculture

The irregular heating of the atmosphere has significant consequences for aquatic environments:

- **Temperature Gradients in Water:** Uneven atmospheric heating leads to temperature differences in surface waters across latitudes and between coastal and offshore areas. This affects the distribution, physiology, and breeding cycles of aquatic species.
- **Wind Patterns:** Temperature differences in the atmosphere create pressure gradients, which drive wind. Winds, in turn, can cause mixing in water bodies, influence surface currents, and affect upwelling (bringing nutrient-rich deep water to the surface), which is crucial for marine productivity.
- **Precipitation Patterns:** Irregular heating influences global air circulation patterns, which determine where it rains or snows. This affects freshwater availability, river discharge into coastal areas, and salinity levels in estuaries, all vital for certain fisheries and aquaculture.

- **Climate Zones:** The large-scale patterns of irregular heating are fundamental to the development of different climate zones, which dictate the types of aquatic ecosystems and the species that can thrive in them.

In essence, the fact that the atmosphere is not heated evenly sets the stage for a dynamic planet with diverse weather patterns and a wide range of aquatic habitats. Understanding the causes and consequences of this irregular heating is crucial for comprehending the environmental context in which fisheries and aquaculture operate.

Insolation: The Sun's Energy and Aquatic Life

1. **Insolation refers to:**

   A. Heat loss from the Earth
   B. Solar radiation received on Earth's surface
   C. Heat energy stored in oceans
   D. Light from the Moon

2. **The ultimate source of energy for aquatic ecosystems is:**

   A. Earth's core
   B. Wind energy
   C. The Sun
   D. Tidal forces

3. **Which factor affects the intensity of insolation?**

   A. Earth's rotation speed
   B. Cloud type
   C. Angle of solar incidence
   D. Atmospheric pressure

4. **Maximum solar radiation is received at:**

   A. Poles
   B. Equator
   C. Mid-latitudes
   D. Oceans only

5. **Which component of solar radiation is most effective for photosynthesis?**

   A. Ultraviolet
   B. Infrared
   C. Visible light
   D. Microwave

6. **Insolation varies due to:**

   A. Earth's tilt and orbit
   B. Moon phases
   C. Ocean salinity
   D. Tectonic activity

7. **Day length affects insolation by influencing:**

   A. Temperature only
   B. Duration of sunlight received
   C. Tidal movements
   D. Water pH

8. **Insolation is measured in:**

   A. Lux
   B. Joules
   C. Watts per square meter
   D. Degree Celsius

9. Which region has the least annual variation in insolation?

A. Arctic
B. Temperate
C. Tropical
D. Desert

10. What percentage of solar radiation is absorbed by the Earth's surface?

A. 10%
B. 23%
C. 51%
D. 70%

11. Sunlight is vital in aquatic ecosystems for:

A. Fish migration
B. Thermal stratification
C. Photosynthesis by phytoplankton
D. Sediment transport

12. Primary producers in aquatic food chains rely on:

A. Wind energy
B. Chemosynthesis
C. Solar radiation
D. Ocean currents

13. Aquatic life depends on solar energy for:

A. Oxygen generation
B. Cooling effect
C. Water expansion
D. Pressure regulation

14. Solar radiation controls the following in water bodies:

A. Salinity
B. Tidal range
C. Heat budget and productivity
D. Wave energy

15. Photosynthetically Active Radiation (PAR) refers to light in the range of:

A. 100–400 nm
B. 400–700 nm
C. 700–1000 nm
D. Above 1000 nm

16. High insolation leads to:

A. Decreased photosynthesis
B. Increased primary productivity
C. Reduced oxygen solubility
D. Both B and C

17. Excessive solar heating of surface waters can cause:

A. Enhanced mixing
B. Thermal stratification
C. Fish growth
D. Increased oxygen supply

18. Turbid water bodies limit insolation penetration due to:

A. Alkalinity
B. Particulate matter
C. High pH
D. Salt content

19. Deep aquatic zones with no light are called:

A. Littoral zones
B. Aphotic zones
C. Pelagic zones
D. Epilimnion

20. Light availability directly affects the:

A. Water pressure
B. Fish behavior
C. Distribution of aquatic plants
D. Groundwater quality

21. Water temperature is primarily regulated by:

A. Groundwater discharge
B. Sun's energy
C. River input
D. Sediment color

22. Temperature stratification typically occurs in:

A. Shallow ponds
B. Fast-flowing rivers
C. Deep lakes
D. Estuarine backwaters

23. Increased solar radiation leads to a rise in:

A. Salinity
B. Wind velocity
C. Surface water temperature
D. Cloud formation

24. Thermal stratification reduces:

A. Algal growth
B. Vertical mixing of nutrients
C. Evaporation
D. Temperature gradient

25. A sudden decrease in solar radiation can cause:

A. Water cooling
B. Algal bloom
C. Increased DO
D. None of these

26. Solar heating contributes to:

A. Nighttime fish feeding
B. Water column instability
C. Seasonal mixing
D. Diurnal variation in oxygen

27. Sunlight triggers which of the following in aquatic life?

A. Photosynthesis
B. Breeding cycles
C. Circadian rhythms
D. All of the above

28. Sunlight penetrates deeper in:

A. Clear freshwater
B. Turbid estuaries
C. Muddy rivers
D. Algae-rich ponds

29. Diurnal heating of surface waters affects:

A. Bottom feeders
B. Surface-dwelling organisms
C. Only phytoplankton
D. None of these

30. Which organisms are most sensitive to low light conditions?

A. Carnivorous fish
B. Phytoplankton
C. Zooplankton
D. Crabs

31. Seasonal insolation variation affects:

A. Aquatic productivity
B. Nutrient leaching
C. pH changes
D. Groundwater

32. Insolation is highest during:

A. Monsoon
B. Winter
C. Summer
D. Spring

33. In the winter, reduced insolation leads to:

A. High surface temperature
B. Low biological productivity
C. Algal bloom
D. Evaporation increase

34. Which region shows greatest seasonal insolation variation?

A. Equator
B. Tropics
C. Polar regions
D. Coastal belts

35. Cloud cover during monsoon:

A. Increases solar radiation
B. Enhances PAR
C. Reduces insolation
D. Stabilizes water temperature

36. Latitudinal variation in insolation is mainly due to:

A. Ocean current
B. Angle of sun rays
C. Wind direction
D. Human activities

37. Which season supports maximum aquatic primary productivity in India?

A. Post-monsoon
B. Winter
C. Summer
D. Pre-monsoon

38. Solar energy surplus near equator contributes to:

A. Ocean upwelling
B. Strong surface currents
C. Coral bleaching
D. Low productivity

39. Which factor least affects regional insolation?

A. Latitude
B. Altitude
C. Soil texture
D. Cloud cover

40. Upwelling zones benefit from:

A. Warm surface water
B. Deep nutrient-rich water and moderate sunlight
C. Low salinity
D. Stagnant water

41. Solar energy use in aquaculture includes:

A. Water purification
B. Aeration and lighting
C. Feed processing
D. All of the above

42. Shading aquatic ponds reduces:

A. Predator access
B. Solar heating
C. Rainwater input
D. Plant growth

43. Floating solar panels can benefit aquaculture by:

A. Enhancing biodiversity
B. Providing shade and energy
C. Reducing fish yield
D. Polluting the pond

44. Light sensors in aquaculture help monitor:

A. Water depth
B. Chloride levels
C. PAR levels
D. Air humidity

45. Proper insolation management improves:

A. Water drainage
B. Fish coloration
C. Productivity and health of fish
D. Groundwater recharge

46. Artificial lighting in indoor aquaculture is used to simulate:

A. pH variation
B. Natural light cycles
C. Wind stress
D. Pressure gradients

47. Aquatic plant growth can be enhanced by managing:

A. Water hardness
B. Light availability
C. CO_2 exhaust
D. Flow velocity

48. Which color of water surface reflects most solar radiation?

A. Black
B. Blue
C. Green
D. White

49. In solar-powered hatcheries, light intensity should be:

A. Too high
B. Carefully regulated
C. Neglected
D. Pulsed randomly

50. Managing sunlight exposure is key for:

A. Cooling fish
B. Enhancing digestion
C. Healthy photosynthesis and thermal balance
D. Generating rainfall

Answer Key (s)

1	b	2	c	3	c	4	b	5	c	6	a	7	b	8	c	9	c	10	c
11	c	12	c	13	a	14	c	15	b	16	d	17	b	18	b	19	b	20	c
21	b	22	c	23	c	24	b	25	a	26	d	27	d	28	a	29	b	30	b
31	a	32	c	33	b	34	c	35	c	36	b	37	a	38	b	39	c	40	b
41	d	42	b	43	b	44	c	45	c	46	b	47	b	48	d	49	b	50	c

3

Temperature: Factor in Aquatic Ecosystems

Temperature is a fundamental meteorological element that exerts a pervasive influence on all aspects of life, particularly within the aquatic environments you are dedicated to understanding. For fisheries and aquaculture students, comprehending the dynamics of temperature in both the atmosphere and the water is absolutely crucial. In meteorology , temperature means the temperature of air unless it is mentioned otherwise.

Think about the last time you felt the coolness of a mountain stream or the warmth of a shallow pond under the summer sun. These temperature differences are not just sensory experiences; they are critical environmental cues and constraints for fish and other aquatic organisms. Temperature directly affects their metabolic rates, oxygen requirements, growth, reproduction, and even their geographical distribution.

This chapter will delve into the concept of temperature from a meteorological perspective, bridging the gap between atmospheric conditions and their direct impact on the thermal regimes of aquatic habitats. We will explore:

- **What temperature is** and how it is measured in the context of both air and water.
- **The primary drivers of temperature variations**, including solar radiation (insolation), the Earth's surface properties, and the movement of air and water masses.
- **Diurnal and seasonal temperature cycles** and how these fluctuations influence aquatic environments.
- **The crucial relationship between air temperature and water temperature**, examining how heat is exchanged between the atmosphere and water bodies.
- **The concept of thermal stratification** in lakes and oceans, a phenomenon largely driven by temperature differences and with significant implications for water quality and habitat suitability.

- **The profound effects of temperature on aquatic life**, ranging from physiological processes to species distribution and community structure.

By understanding the principles of temperature dynamics discussed in this chapter, you will gain a vital tool for analyzing and predicting the conditions within aquatic ecosystems relevant to both wild fisheries management and aquaculture practices. Whether you are considering the optimal temperature for culturing a specific fish species or assessing the impact of climate change on fish habitats, a solid grasp of temperature and its controls is indispensable.

3.1 Measurement of Temperature

Thermometers are used to measure temperature. The maximum thermometer measures the highest temperature during the day while minimum thermometer is used to record the lowest temperature recorded during night.

3.1.1 Maximum Thermometer

The maximum thermometer is a type of mercury-in-glass thermometer designed to measure the maximum temperature reached during a specific period. Its construction includes:

- **Mercury-filled bulb**: A bulb containing mercury, which expands when heated.
- **Capillary tube**: A narrow tube connected to the bulb, allowing mercury to rise as it expands.
- **Constriction**: A narrow point in the capillary tube that prevents the mercury column from falling back when the temperature drops.
- **Scale**: A calibrated scale marked with temperature values, allowing users to read the maximum temperature.
- **Working principle:** The maximum thermometer works on the principle of thermal expansion:
- **Temperature rise**: As the temperature increases, the mercury in the bulb expands, causing the mercury column to rise in the capillary tube.
- **Constriction**: When the temperature reaches its maximum, the mercury column is pushed past the constriction in the capillary tube.
- **Temperature drop**: As the temperature drops, the mercury column breaks at the constriction, trapping the mercury above it.
- **Maximum temperature reading**: The maximum temperature is indicated by the top of the mercury column.

- **Exposure**: To ensure accurate measurements, the maximum thermometer is typically:
- **Mounted on a wooden board**: The thermometer is fixed to a wooden board, providing a stable base and kept horizontally in a Stevenson's screen.
- **Bulb exposed**: The bulb is positioned outside the wooden board, allowing it to be exposed to the air.
- **Guarded by a U-shaped metal frame**: A U-shaped metal frame protects the thermometer from damage while allowing for good airflow around the bulb.
- **Shaded**: Often placed in Stevenson screen, to protect it from direct sunlight.
- **Well-ventilated**: Installed in a well-ventilated area to allow for free airflow.
- **Method of Taking Observati**ons: To take observations with a maximum thermometer:
- **Reset**: Before taking a reading, the thermometer needs to be reset by whirling or shaking it to bring the mercury column back down to the bulb.
- **Read**: Read the maximum temperature indicated by the top of the mercury column.The reading of the thermometer indicates the maximum temperature attained by the air during the day time.
- **Record**: Record the maximum temperature reading.
- **Reset again**: Reset the thermometer again after taking the reading to prepare it for the next observation.

By following these steps and ensuring proper exposure, maximum thermometers provide accurate measurements of the highest temperature reached during a specific period, which is essential for meteorological observations.

3.1.2 Stevenson's Screen

- **Single Stevenson Screen:** A single Stevenson screen is a wooden, louvered enclosure used to house meteorological instruments, such as thermometers and hygrometers. Its design allows for airflow while protecting the instruments from direct sunlight and precipitation. The standard dimensions of a single Stevenson screen are:
  - **Height**: 122 cm (4 feet) above the ground
  - **Width**: 76 cm (30 inches)y

  - **Depth**: 56 cm (22 inches)
- **Construction** : The screen is typically constructed from wood and features:
  - **Louvered sides**: Slatted sides that provide ventilation and help to reduce the impact of radiation on the instruments
  - **Double roof**: Not always a feature of single screen, but often a roof with sufficient overhang to protect from direct sun and rain
  - **Door**: A door at the north side (in the northern hemisphere) that allows for easy access to the instruments, facing away from the direct sun
  - It is painted white to reflect solar radiation incident upon it.
- **Features:** The Stevenson screen is designed to:
  - **Provide shade**: Protect instruments from direct sunlight
  - **Allow ventilation**: Enable airflow around the instruments to ensure accurate readings
  - **Protect from precipitation**: Shield instruments from rain and other forms of precipitation
- **Installation :** The screen should be installed in an open area, away from buildings and trees, and at a standard height above the ground.

The Stevenson screen plays a crucial role in ensuring accurate and reliable meteorological measurements.

• Double Stevenson's Screen

A **double Stevenson's screen** is a specialized version of the standard Stevenson screen, designed to provide even better insulation and radiation shielding for meteorological instruments. While the principles of construction, installation, and use are similar to a single screen, the double screen has enhanced features.

- **Construction::** A double Stevenson's screen is essentially a larger box with double-layered walls, roof, and often a double floor. The construction typically involves:
  - **Material:** Traditionally made of wood (often seasoned teak) which is a poor conductor of heat. Modern screens might use other materials like certain polymers or aluminum with insulating properties.
  - **Double Louvered Sides:** The sides have two sets of louvers, an outer and an inner layer, with an air gap between them. These louvers are angled downwards and outwards to allow free airflow while preventing direct sunlight, rain, and snow from entering. The double layer provides enhanced shielding from radiation.

- **Double Roof:** A double-layered roof with an air gap in between provides insulation from direct solar heating. The outer roof reflects solar radiation, and the air gap acts as an insulator. The roof is typically sloped to allow rainwater to run off.
- **Floor:** The floor is often made of overlapping boards with air spaces, sometimes also double-layered, to further insulate the interior from ground radiation.
- **White Paint:** The entire screen is painted white to maximize the reflection of solar radiation, preventing the inside from heating up due to absorbed sunlight.
- **Hinged Door:** A hinged door on one side (typically facing north in the Northern Hemisphere and south in the Southern Hemisphere) allows access to the instruments. It usually has a mechanism (like chains) to hold it open horizontally.
- **Frame/Stand:** The screen is mounted on a sturdy stand, usually made of wood or metal, raising it above the ground.

The "double" aspect refers primarily to the double-layered louvered sides, and often the double roof, providing superior insulation and radiation protection compared to a single-walled screen. Some larger "double" Stevenson screens might also be wider to accommodate more instruments.

- **Dimensions:** The dimensions of a double Stevenson's screen (often referred to as a "large" Stevenson screen) are typically larger than a single screen to house a wider array of instruments, including recording types. Approximate internal dimensions for a larger/double screen can be around:
  - **Length:**~1325mm
  - **Width:**~315mm
  - **Height:**~412mm

However, these dimensions can vary depending on the manufacturer and specific design. The height when installed on a stand is crucial, with the thermometer bulbs ideally positioned about 1.25 to 2 meters (around 4 feet) above the ground.

- **Installation:** Proper installation is crucial for the accuracy of the measurements:
  - **Location:** The screen should be located in an open area, away from obstructions like buildings and trees (ideally at a distance of at least twice the height of the obstruction). A level, grassy surface is preferred to represent typical ground conditions.

- **Orientation:** The door should face north in the Northern Hemisphere (and south in the Southern Hemisphere) to prevent direct sunlight from falling on the instruments when the door is opened during observations.
- **Leveling:** The base of the stand must be level to ensure the instruments inside are also correctly oriented.
- **Height:** The screen should be mounted such that the sensing parts (e.g., bulbs of thermometers) of the instruments inside are approximately 1.25 meters (UK standard) above the ground. This minimizes the influence of ground-level temperature gradients.
- **Security:** In some settings, the screen might be secured to prevent tampering.

- **Use:** The primary use of a double Stevenson's screen is to house meteorological instruments, shielding them from:
  - **Direct solar radiation:** Prevents the instruments from being heated by the sun, which would give a false reading of the air temperature.
  - **Precipitation:** Protects delicate instruments from rain, snow, and hail.
  - **Wind:** Minimizes the direct impact of wind on the instruments, which can affect temperature readings, especially for wet-bulb thermometers.
  - **Ground radiation:** The raised platform and insulated floor help to minimize the influence of heat radiating from the ground.

Instruments commonly housed in a double Stevenson's screen include

- **Thermograph and Hygrograph:** Recording instruments that continuously measure and record temperature and humidity over time.
- Sometimes, a **barometer or barograph** (for recording pressure) might also be placed in a larger screen.

By providing a standardized shelter, the double Stevenson's screen ensures that measurements taken at different locations and times are comparable, as the instruments are all exposed to the ambient air under similar shielded conditions. Regular maintenance, such as cleaning and repainting (to maintain the high reflectivity of the white paint), is important for the screen to function effectively.

3.1.3 Minimum Thermometer

A **minimum thermometer** used in a meteorological observatory is typically an **alcohol-in-glass thermometer with an index**, designed to record the lowest temperature reached over a specific period.

- **Construction**
  - **Glass Tube:** A glass tube contains alcohol in a bulb at one end. Alcohol's low freezing point (around −114 °C) makes it suitable for measuring meteorological minimum temperatures. The tube has a fine bore.
  - **Index:** A small, elongated glass or enamel index resides within the bore, free to move with the alcohol column but resistant to returning on its own due to surface tension or a slight constriction.
  - **Scale:** A temperature scale, usually in Celsius, is marked alongside the glass tube.
- **Working**
  - **Falling Temperature:** As the temperature drops, the alcohol contracts, and the meniscus moves towards the bulb, dragging the index with it due to surface tension.
  - **Minimum Reached:** At the minimum temperature, the alcohol stops contracting, and the meniscus halts, leaving the index at the lowest point.
  - **Rising Temperature:** When the temperature rises, the alcohol expands and moves up the bore, but the index remains at the minimum temperature position.
  - **Reading:** The minimum temperature is read from the end of the index furthest from the bulb.
  - **Resetting:** The thermometer is reset by holding it vertically, bulb upwards, and gently tapping it, allowing the index to slide back to the alcohol meniscus.
- **Installation:** In a Stevenson screen, the minimum thermometer is often mounted on a **wooden board**. The typical installation involves:
  - **Mounting on a Wooden Board:** The thermometer is affixed to a wooden board, often alongside other thermometers.
  - **Bulb Exposure:** The bulb of the thermometer extends outwards, away from the wooden board, allowing it to be freely exposed to the air within the screen.

- **Protection with a U-shaped Metal Frame:** A **U-shaped metal frame** is fixed to the wooden board and positioned to protect the glass tube of the thermometer from accidental damage, without obstructing airflow around the bulb or hindering the movement of the index.
- **Orientation:** The board with the mounted thermometer is typically placed horizontally, or with a slight downward tilt towards the bulb, on supports inside the Stevenson screen.
- **Height:** The bulb of the minimum thermometer should be positioned approximately 1.25 meters above the ground, consistent with the standard height for meteorological temperature measurements.

This mounting arrangement ensures the thermometer is securely held, protected from minor impacts, and that its bulb is exposed to the ambient air within the Stevenson screen to accurately measure the minimum temperature.

3.2 Periodic, Horizontal and Vertical Variations of Temperature

Let's explore the variations in temperature in detail, considering periodic, horizontal, and vertical aspects, and their relevance to fisheries and aquaculture.

3.2.1 Periodic Variations of Temperature

These are cyclical changes in temperature that occur over time. The most significant periodic variations relevant to aquatic environments are:

- **Diurnal (Daily) Variations**
  - This refers to the temperature changes within a 24-hour cycle.
  - **Atmosphere:** Typically, the warmest air temperatures occur in the afternoon (after the time of maximum solar radiation) due to the lag in heating, and the coldest temperatures occur around sunrise.
  - **Aquatic Environments:** Water has a higher heat capacity than air and land, so it heats up and cools down more slowly. Diurnal temperature variations in larger, deeper water bodies are less pronounced than in shallow ponds or the air above them. However, shallow aquaculture ponds can experience significant daily temperature swings, affecting the metabolism and stress levels of cultured organisms. Coastal intertidal zones also experience strong diurnal temperature changes as they are exposed and submerged with tides.
- **Seasonal (Annual) Variations**
  - These are the temperature changes throughout the year, driven by the Earth's revolution around the sun and the tilt of its axis

- **Atmosphere:** We experience distinct seasons with varying average temperatures depending on latitude.
- **Aquatic Environments:** Water temperatures follow seasonal patterns, lagging behind air temperatures. Lakes and oceans in temperate regions will have warmer surface waters in summer and colder waters in winter. These seasonal changes are critical cues for the life cycles of many aquatic species, influencing reproduction, migration, and growth. Aquaculture operations in open systems are directly influenced by these seasonal temperature shifts, often dictating which species can be cultured and the timing of production cycles.

- **Longer-Term Periodic Variations:**
  - While less predictable and often overlapping with climate change, there can be longer-term cycles in temperature (e.g., related to solar activity or ocean oscillations). These can have subtle but cumulative effects on aquatic ecosystems over years to decades.

3.2.2 Horizontal Variations of Temperature

These refer to the temperature differences observed across different geographical locations on the Earth's surface (at roughly the same altitude).

- **Latitude:** This is the most significant factor. Temperatures generally decrease from the equator towards the poles due to the angle at which solar radiation strikes the Earth surface .This broad pattern dictates the major thermal zones of the world and the types of aquatic ecosystems found in each (e.g., tropical coral reefs vs. polar seas). This directly influences the natural distribution of fish species suitable for different latitudes and the potential for aquaculture of specific species in different regions.
- **Land and Water Distribution:** Water moderates temperature. Coastal areas have smaller temperature ranges compared to inland areas at the same latitude. This maritime influence is crucial for coastal fisheries and aquaculture, often providing more stable thermal environments.
- **Ocean Currents:** Warm currents transport heat towards higher latitudes, while cold currents transport cooler water towards lower latitudes, significantly affecting the temperatures of adjacent coastal waters and influencing the distribution of marine species. For example, the Gulf Stream warms the North Atlantic, while the cold Humboldt Current cools the Pacific coast of South America, both with profound impacts on local fisheries.

- **Altitude:** While technically a vertical variation, when comparing different locations, higher altitudes generally have lower temperatures. This affects freshwater ecosystems; for instance, higher elevation streams and lakes tend to be colder, often supporting different fish species than warmer, lower elevation waters.

3.2.3 Vertical Variations of Temperature

These describe how temperature changes with altitude in the atmosphere or with depth in water bodies.

- **Atmosphere:** In the troposphere (the lowest layer), temperature generally decreases with increasing altitude (Fig. 3.1). This is known as the normal lapse rate (about 6.5°C per kilometer). However, under certain conditions, temperature inversions can occur where temperature increases with height, which can affect air pollution and local weather patterns that might indirectly impact nearby aquatic systems (e.g., through altered precipitation).
- **Aquatic Environments:** Vertical temperature variations in water bodies are critical:
  - **Surface waters** are typically warmed by solar radiation.
  - **Deeper waters** are generally colder as sunlight penetration decreases with depth.
  - **Thermal Stratification:** In lakes and oceans, especially during warmer months, this can lead to the formation of distinct layers (epilimnion - warm surface layer, thermocline - zone of rapid temperature change, hypolimnion - cold deep layer). Stratification affects dissolved oxygen levels, nutrient distribution, and the vertical distribution of aquatic organisms. For aquaculture in deep water systems (like some net pens), understanding these vertical temperature gradients is important.
  - In shallow water bodies, vertical temperature differences might be minimal due to mixing.
  - In oceans, the vertical temperature profile also shows a decrease with depth, but the structure can be more complex with a surface mixed layer, a thermocline, and deep, cold water. This influences the distribution of pelagic fish species. Note that in very deep ocean waters, the temperature is quite uniform and cold.

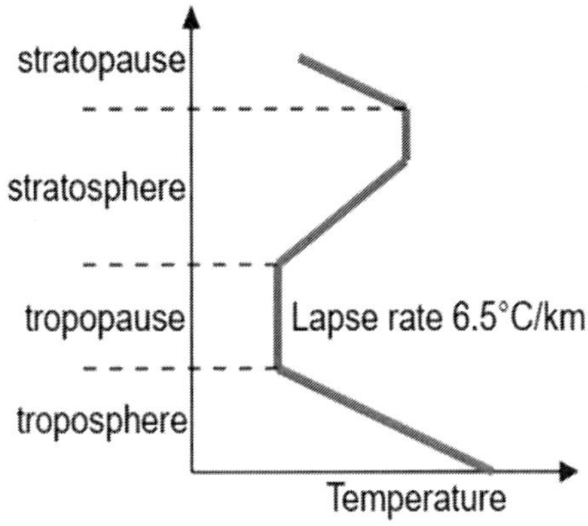

Fig. 3.1: Lapse rate

3.2.4 Relevance to Fisheries and Aquaculture

Understanding these periodic, horizontal, and vertical temperature variations is vital for:

- **Species Distribution:** Knowing the thermal tolerances of different fish and shellfish helps predict where they naturally occur and where they can be successfully farmed.
- **Physiological Processes:** Temperature directly affects metabolic rates, growth, reproduction, and immune responses of aquatic organisms.
- **Water Quality:** Temperature influences dissolved oxygen levels, stratification, and the solubility of other substances in water.
- **Site Selection for Aquaculture:** Choosing locations with appropriate temperature regimes that match the requirements of the cultured species.
- **Climate Change Impacts:** Predicting how changes in these temperature patterns will affect aquatic habitats and the distribution and productivity of fisheries resources.

3.3 Effect of Vertical Motions on Temperature

Vertical motions in the atmosphere, whether upward (ascent) or downward (subsidence), play a significant role in changing air temperature (Fig. 3.2). The primary reason for this temperature change is **adiabatic processes**.

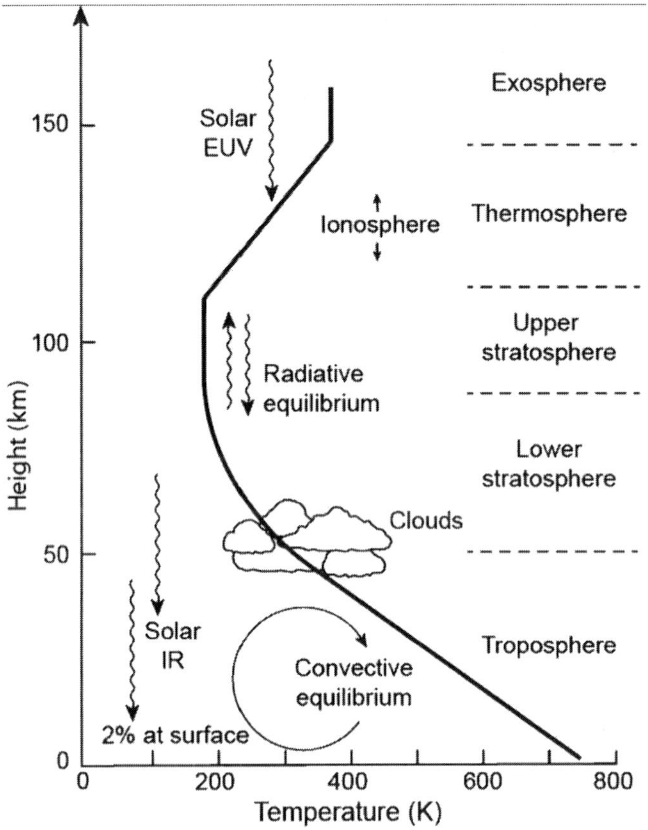

Fig. 3.2: Vertical profile of air temperature in the atmosphere

3.3.1 Adiabatic Processes

An adiabatic process is one in which a parcel of air rises or sinks without exchanging heat with its surroundings. The temperature changes solely due to the expansion or compression of the air parcel as it moves to areas of different pressure.

- **Rising Air (Expansion):** As an air parcel rises, it moves into regions of lower atmospheric pressure. This lower pressure allows the air parcel to expand. When the air expands, its molecules spread out, and the internal energy of the parcel decreases, resulting in a **decrease in temperature**.
- **Sinking Air (Compression):** Conversely, as an air parcel sinks, it moves into regions of higher atmospheric pressure. This higher pressure compresses the air parcel, causing the air molecules to collide more frequently. It This increases the internal energy of the parcel, resulting in an **increase in temperature**.

3.3.2 Lapse Rates

The rate at which temperature changes with altitude is called the **lapse rate**. There are two key adiabatic lapse rates:

- **Dry Adiabatic Lapse Rate (DALR):** This applies to unsaturated air (air with a relative humidity below 100%). The DALR is approximately 9.8 °C per kilometer (or 5.5 °F per 1000 feet). So, for every kilometer an unsaturated air parcel rises, its temperature will decrease by about 9.8 °C, and for every kilometer it sinks, its temperature will increase by the same amount.
- **Moist (or Saturated) Adiabatic Lapse Rate (MALR):** This applies to saturated air (air with a relative humidity of 100%, where condensation is occurring). As saturated air rises and cools, water vapor condenses, releasing latent heat. This release of heat partially offsets the cooling due to expansion, so the MALR is less than the DALR. The MALR is not constant and varies with temperature and pressure but is typically in the range of 4–9 °C per kilometer.

3.3.3 Effects on Atmospheric Stability

The effect of vertical motion on temperature is crucial for determining the **stability of the atmosphere**:

- If a rising air parcel becomes colder than the surrounding air, it will tend to sink back to its original level, indicating a **stable atmosphere**, which inhibits vertical motion.
- If a rising air parcel remains warmer than the surrounding air, it will continue to rise (due to buoyancy), indicating an **unstable atmosphere**, which promotes vertical motion and can lead to cloud development and storms.

3.3.4 Relevance to Fisheries and Aquaculture (Indirect)

The direct effect of these adiabatic temperature changes is on the atmosphere. However, these atmospheric temperature changes driven by vertical motions can indirectly influence aquatic environments:

- **Cloud Formation and Solar Radiation:** Upward motion (ascent) often leads to cooling and condensation, resulting in cloud formation. Clouds can reduce the amount of solar radiation (insolation) reaching the surface of water bodies, thus affecting water temperature and primary productivity.

- **Precipitation:** Vertical motions are essential for precipitation. Rising air cools, leading to condensation and potentially rainfall or snowfall, which directly impacts freshwater systems (river flow, lake levels) and coastal salinity.
- **Wind Patterns:** Vertical motions are part of larger atmospheric circulation patterns that drive winds. Winds can affect surface water mixing, upwelling in oceans, and the temperature of shallow aquaculture ponds.
- **Extreme Weather:** Strong vertical motions are associated with thunderstorms and other severe weather events that can directly impact aquaculture infrastructure (e.g., damage from high winds or flooding from heavy rain) and disrupt fishing activities.

In summary, vertical motions in the atmosphere cause temperature changes through adiabatic processes, which in turn influence atmospheric stability, cloud formation, precipitation, and wind patterns.These atmospheric conditions then have significant direct and indirect effects on aquatic environments relevant to fisheries and aquaculture.

Temperature-Factor in Aquatic Ecosystems

1. **Which parameter most directly controls the metabolic rate of aquatic organisms?**
   A. pH B. Temperature
   C. Salinity D. Light intensity
2. **Temperature in aquatic systems primarily comes from:**
   A. Geothermal sources B. The Moon
   C. Solar radiation D. Volcanic activity
3. **The unit commonly used to measure water temperature is:**
   A. Pascal B. Joule
   C. Degree Celsius D. Kelvin
4. **Which layer in a thermally stratified lake is warmest?**
   A. Epilimnion B. Metalimnion
   C. Hypolimnion D. Benthos
5. **Sudden changes in water temperature can cause:**
   A. Increased pH B. Algal bloom
   C. Fish stress or death D. Salinity rise

6. **Temperature in aquatic ecosystems affects:**
   A. Reproduction
   B. Growth rate
   C. Oxygen solubility
   D. All of the above
7. **Thermal pollution refers to:**
   A. Noise in water
   B. Heat released into water bodies from industries
   C. Temperature drop due to rainfall
   D. Natural geothermal heat
8. **Thermal stratification typically occurs in:**
   A. Fast-flowing rivers
   B. Shallow tanks
   C. Deep, still lakes
   D. Rain-fed ponds
9. **Aquatic organisms are generally classified as:**
   A. Poikilotherms
   B. Homeotherms
   C. Thermophiles
   D. Psychrophiles
10. **Fish body temperature usually:**
   A. Remains constant
   B. Increases in cold water
   C. Fluctuates with water temperature
   D. Is higher than air temperature
11. **Temperature affects dissolved oxygen because:**
   A. Oxygen dissolves better in warm water
   B. Cold water holds more oxygen
   C. Oxygen is independent of temperature
   D. Warm water has no effect on oxygen
12. **Optimum temperature range for most warm-water fish is:**
   A. 5–10°C
   B. 10–15°C
   C. 20–30°C
   D. 35–45°C
13. **Beyond their thermal tolerance limit, fish may experience:**
   A. Enhanced growth
   B. Lethargy and mortality
   C. More feeding
   D. Spawning
14. **Temperature influences the solubility of:**
   A. Heavy metals only
   B. Gases like oxygen and CO_2
   C. Sediments
   D. Salts only
15. **Which organism group is most sensitive to thermal fluctuations?**
   A. Amphibians
   B. Phytoplankton
   C. Coral reefs
   D. Molluscs
16. **The process of oxygen depletion in warmer water is called:**
   A. Eutrophication
   B. Thermal decay
   C. Hypoxia
   D. Oxidation

17. Temperature increase causes fish respiration rate to:

A. Decrease
B. Remain constant
C. Increase
D. Stop

18. Thermal stress can delay:

A. Photosynthesis
B. Spawning in fish
C. Sediment transport
D. Plankton movement

19. Cold-blooded animals are also known as:

A. Endotherms
B. Poikilotherms
C. Thermophiles
D. Homeotherms

20. Temperature changes cause vertical movement in fish for:

A. Avoiding light
B. Avoiding predators
C. Seeking thermal comfort
D. Escaping nutrients

21. Stratification is most stable during:

A. Winter
B. Monsoon
C. Summer
D. Spring

22. Layer with greatest temperature change in a stratified lake is:

A. Epilimnion
B. Metalimnion
C. Hypolimnion
D. Benthic zone

23. Spring and autumn overturn in lakes occur due to:

A. Strong wind
B. Earthquakes
C. Uniform water temperature
D. Evaporation

24. In winter, surface water in lakes:

A. Is the warmest layer
B. Freezes and becomes less dense
C. Has high salinity
D. Is full of nutrients

25. Which layer contains least oxygen during stratification?

A. Surface
B. Thermocline
C. Hypolimnion
D. Epilimnion

26. Mixing of water during overturn helps to:

A. Stratify the lake
B. Increase oxygen and nutrient availability
C. Reduce fish spawning
D. Lower light penetration

27. Thermal stratification disappears in:

A. Still ponds
B. Large lakes
C. Mixing seasons (spring/autumn)
D. Summer

28. Thermal layers affect fish distribution because of:

A. Light variation B. Temperature preference
C. Predator avoidance D. Substrate type

29. Shallow ponds are less likely to exhibit:

A. Thermal pollution B. Thermal layering
C. Diurnal heating D. Microbial growth

30. The upper mixed layer of water is called:

A. Thermocline B. Epilimnion
C. Hypolimnion D. Metalimnion

31. A common source of anthropogenic heat in aquatic bodies is:

A. Rainfall B. Hydropower
C. Industrial effluents D. Aquatic weeds

32. Climate change may lead to:

A. Constant water temperature B. Stable thermal layers
C. Increased thermal stress on aquatic life D. Lower evaporation

33. Rising global temperatures may cause:

A. Higher DO levels B. Coral bleaching
C. Reduced fish breeding D. Both B and C

34. Thermal discharge from power plants may lead to:

A. Hypersalinity B. Thermal enrichment
C. Local extinction of species D. Genetic variation

35. Sustainable temperature management involves:

A. Ignoring thermal waste B. Releasing hot water directly
C. Cooling waste water before discharge D. Mixing hot and cold water

36. Fish farmers must monitor temperature because:

A. Fish avoid hot water B. Feed efficiency depends on temperature
C. Disease outbreaks are temperature-sensitive D. All of the above

37. Climate warming can shift fish populations:

A. Towards poles B. Towards equator
C. Deeper in lakes D. Into estuaries

38. Coral bleaching is caused by:

A. Salinity increase B. High temperature stress
C. Increased oxygen D. Predator invasion

39. Fish farming tanks should maintain temperature by:

A. Using heat pumps
B. Shading
C. Controlled aeration
D. All of the above

40. Thermal refuges are:

A. Places of high salinity
B. Cool zones offering relief from heat stress
C. Heated water areas
D. Wind-exposed surfaces

41. Fish adapt to temperature by adjusting:

A. Morphology
B. Behavioral patterns
C. Enzyme activity
D. All of the above

42. Which tool is commonly used to monitor water temperature?

A. Spectrometer
B. Thermometer
C. Hydrometer
D. Salinometer

43. Sudden cold shock may result in:

A. Higher feeding rate
B. Disease resistance
C. Fish mortality
D. Photosynthesis boost

44. Acclimatization in fish refers to:

A. Migration
B. Genetic change
C. Physiological adjustment to temperature
D. Behavioral reproduction

45. Thermal tolerance is best defined as:

A. Water's heating rate
B. Range of temperature an organism can survive
C. Salinity range
D. Oxygen saturation level

46. Cooler water temperatures slow down:

A. Fish metabolism
B. Algal respiration
C. Microbial activity
D. All of the above

47. Long-term exposure to sub-optimal temperature leads to:

A. Acclimatization
B. Stunting or death
C. Reproduction increase
D. Oxygen gain

48. Bioindicators of thermal pollution include:

A. Coral reefs
B. Zooplankton
C. Sensitive fish species
D. All of the above

49. An effective management practice in aquaculture is:

A. Ignoring water temperature
B. Monitoring daily temperature trends
C. Allowing thermal mixing
D. Increasing nutrient input

50. Maintaining optimal temperature range ensures:

A. Disease outbreaks
B. Poor feed conversion
C. High mortality
D. Efficient fish growth and health

Answer Key (s)

1	b	2	c	3	c	4	a	5	c	6	d	7	b	8	c	9	a	10	c
11	b	12	c	13	b	14	b	15	c	16	c	17	c	18	b	19	b	20	c
21	c	22	b	23	c	24	b	25	c	26	b	27	c	28	b	29	b	30	b
31	c	32	c	33	d	34	c	35	c	36	d	37	a	38	b	39	d	40	b
41	d	42	b	43	c	44	c	45	b	46	d	47	b	48	d	49	b	50	d

4

Water Vapor in the Atmosphere and Humidity

The presence of water in its various phases (gas, liquid, and solid) within the atmosphere plays a pivotal role in shaping our weather and climate. This chapter, "Moisture in the Atmosphere," will delve into the crucial concept of atmospheric humidity – the amount of water vapor in the air. Water, unique in its ability to exist in all three states at Earth's temperatures, undergoes phase changes that are fundamental to the formation of clouds, precipitation, and atmospheric stability. Understanding the sources of atmospheric moisture, the ways it is measured and expressed, and its influence on atmospheric processes is essential for comprehending a wide range of meteorological phenomena (Fig. 4.1). We will explore concepts such as evaporation, condensation, humidity indices, and the role of moisture in energy transfer within the Earth-atmosphere system. Join us as we unravel the dynamics of this vital component of our atmospheric environment.

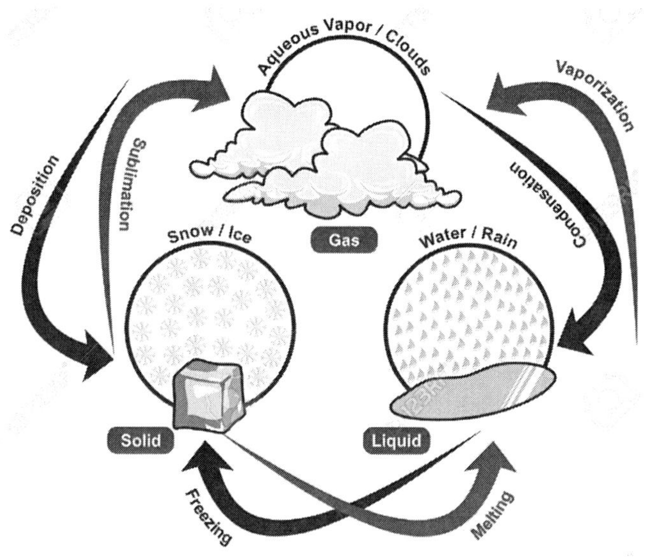

Fig. 4.1: Changes in states of water

4.1 Expressions Used for Water Vapor in the Atmosphere

Let's look into the concepts of vapor pressure, saturated vapor pressure, vapor pressure deficit, and dew point with detailed explanations:

4.1.1 Vapor Pressure (e)

- **Definition:** Vapor pressure is the partial pressure exerted by water vapor molecules in the air. It represents the contribution of water vapor to the total atmospheric pressure. Imagine air as a mixture of gases; the vapor pressure is the "pressure" exerted solely by the water vapor component of that mixture.
- **Detailed Explanation**
  - Water molecules in liquid or solid form (like the surface of a lake or an ice crystal) are constantly in motion. Some of these molecules gain enough kinetic energy to escape into the gaseous phase (evaporation or sublimation).
  - These water vapor molecules then exert their own pressure as they move and collide with surfaces around them. This pressure is the vapor pressure.
  - The amount of water vapor in the air, and consequently the vapor pressure, depends on factors like the availability of water for evaporation/sublimation and the ongoing processes of evaporation/sublimation versus condensation/deposition.

4.1.2 Saturated Vapor Pressure (es)

- **Definition:** Saturated vapor pressure is the maximum possible partial pressure of water vapor that can exist in the air at a given temperature. It represents the point at which the air is holding the maximum amount of water vapor it can before condensation (or deposition) begins.
- **Detailed Explanation**
  - At any given temperature, there's a limit to how much water vapor the air can "hold." This limit is determined by the energy of the air molecules. Warmer air molecules have more energy and can prevent water vapor molecules from easily condensing back into liquid water, thus allowing more water vapor to exist in the gaseous phase.
  - When the rate of evaporation (or sublimation) equals the rate of condensation (or deposition), the air is said to be saturated, and the pressure exerted by the water vapor at this equilibrium is the saturated vapor pressure.

- Saturated vapor pressure is solely a function of temperature. As temperature increases, the saturated vapor pressure also increases non-linearly. This means warmer air can hold significantly more water vapor at saturation than colder air.
- You can think of saturated vapor pressure as the "capacity" of the air to hold water vapor at a specific temperature.
- When the actual vapor pressure equals the saturated vapor pressure ($e=es$), the relative humidity is 100%, and the air is saturated.

4.1.3 Vapor Pressure Deficit (VPD)

- **Definition:** Vapor Pressure Deficit (VPD) is the difference between the saturated vapor pressure (es) and the actual vapor pressure (e) at a given temperature.

  $VPD = es - e$
- **Detailed Explanation**
  - VPD essentially indicates how "thirsty" the air is. A high VPD means the air is far from being saturated and has a high capacity to hold more water vapor, leading to increased evaporation rates from surfaces (like plants, soil, and water bodies).
  - A low VPD means the air is closer to saturation, and the driving force for evaporation is smaller. When VPD is zero ($es=e$), the air is saturated, and the net evaporation rate is zero (equilibrium).
  - VPD is important in various fields, including agriculture (plant transpiration), ecology (evaporation rates), and even wildfire risk assessment (fuel dryness).
  - It's often considered a more direct measure of the evaporative demand of the atmosphere than relative humidity because it directly relates to the pressure difference driving moisture movement.

4.1.4 Dew Point (Td)

- **Definition:** The dew point is the temperature to which air must be cooled at a constant pressure and constant water vapor content for saturation to occur. At this temperature, the air can no longer hold all of its water vapor, and condensation begins to form dew, fog, or clouds.
- **Detailed Explanation**
  - Imagine a parcel of air with a certain amount of water vapor (and thus a specific vapor pressure, e). If you cool this air at a constant pressure without adding or removing any water vapor, the saturated

vapor pressure (es) of the air decreases because es is temperature-dependent.

- The dew point is the temperature at which this cooling causes the saturated vapor pressure (es) to become equal to the actual vapor pressure (e) of the air. At this point, the air becomes saturated (100% relative humidity), and further cooling will lead to condensation.
- A high dew point indicates a large amount of moisture in the air because the air needs to be cooled significantly less to reach saturation. A low dew point indicates drier air.
- The difference between the air temperature and the dew point is a good indicator of how close the air is to saturation. A small difference means the relative humidity is high, and the air is close to saturation. A large difference means the relative humidity is low, and the air is dry.

- **In summary**
  - **Vapor Pressure** tells us how much water vapor is currently in the air.
  - **Saturated Vapor Pressure** tells us the maximum amount of water vapor the air *could* hold at the current temperature.
  - **Vapor Pressure Deficit** tells us the "drying potential" of the air.
  - **Dew Point** tells us the temperature at which condensation would begin if the air were cooled.

4.2 Concept of Saturation of Air

Here's what saturation indicates

- **Maximum Water Vapor Capacity:** Saturated air has reached its limit for holding water vapor at its current temperature (Fig. 4.2). Warmer air can hold more water vapor at saturation than colder air.
- **Equilibrium:** It signifies a dynamic equilibrium between the liquid (or solid) phase of water and its gaseous phase (water vapor). For every water molecule that evaporates, one condenses, resulting in no net change in the amount of water vapor in the air.
- **Relative Humidity of 100%:** When the air is saturated, the actual amount of water vapor present is equal to the maximum amount it can hold. This results in a relative humidity of 100%.
- **Onset of Condensation:** If saturated air is cooled further (without adding or removing water vapor), it will become supersaturated, leading to condensation (formation of liquid water like dew or fog) or deposition

(formation of ice like frost). The temperature at which saturation occurs upon cooling is the dew point (for liquid water) or frost point (for ice).

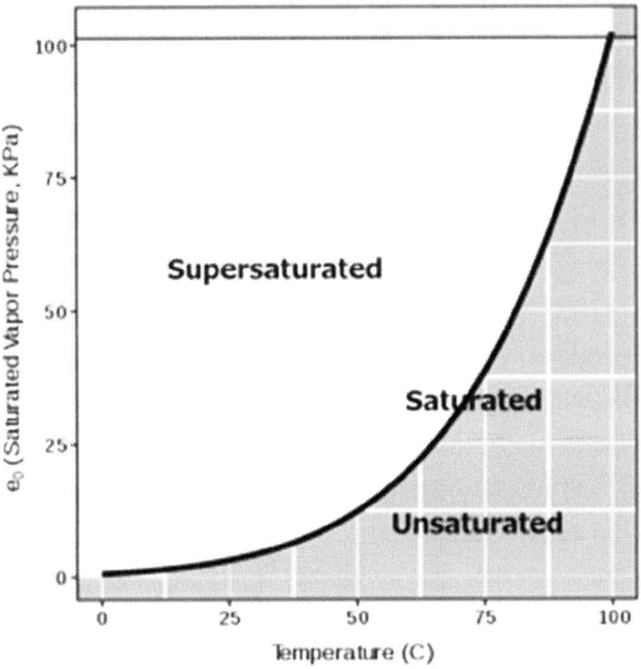

Fig.4.2: Temperature and water vapour relationship

In summary, saturation indicates that the air is "full" of water vapor for its current temperature, and any further addition of moisture or decrease in temperature will likely result in condensation. It's a critical threshold for the formation of clouds and precipitation.

4.2.1 Relative Humidity

Relative humidity (RH) is a measure of the amount of water vapor present in the air compared to the maximum amount of water vapor the1 air could hold at a given temperature. It is expressed as a percentage. Think of it this way: warmer air has a greater capacity to hold water vapor than cooler air. Relative humidity tells you how "full" the air is with water vapor relative to its total capacity at that specific temperature (Fig.4.3).

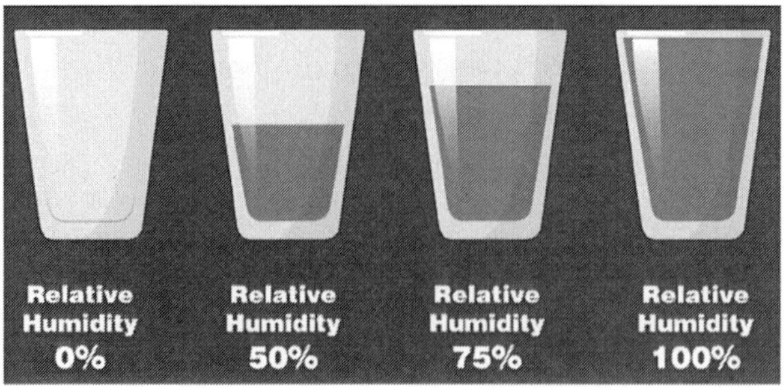

Fig. 4.3: Relative humidity

- **100% RH** means the air is saturated and cannot hold any more water vapor. At this point, condensation (or deposition) is likely to occur, leading to the formation of dew, fog, or clouds.
- **0% RH** means the air is completely dry and contains no water vapor.

• Formula to Calculate Relative Humidity

Relative humidity can be calculated in a few ways, but the most common conceptual formula is:

RH = (Actual amount of water vapor in the air / Maximum amount of water vapor the air can hold at that temperature) * 100%

More formally, using vapor pressures:

RH = (Actual Vapor Pressure (e) / Saturated Vapor Pressure (es)) * 100 %

Where:

- (e) is the partial pressure exerted by the water vapor in the air.
- (e_s) is the maximum vapor pressure that water can exert at that specific temperature (saturated vapor pressure).

There are also more complex formulas that use temperature and dew point to calculate relative humidity, as seen in some of the search results you provided, but the core concept remains the ratio of the actual moisture to the potential moisture.

Based on the current weather in Hyderabad, the humidity is 55%.

- **Significance of Relative Humidity:** Relative humidity is a crucial atmospheric variable for several reasons.
  - **Human Comfort:** It significantly affects how we perceive temperature.6 High relative humidity makes us feel hotter because the evaporation of sweat from our skin is reduced.

- Conversely, low relative humidity can make us feel cooler due to increased evaporation. The "feels like" temperature often incorporates the effect of humidity.
- **Weather Forecasting:** RH is a key indicator of the likelihood of precipitation, dew, and fog. High RH suggests that the air is close to saturation, making condensation more likely with further cooling or the addition of more moisture.
- **Agriculture:** It influences plant transpiration rates and the likelihood of dew formation, which can impact plant health and the spread of diseases.
- **Material Science and Preservation:** RH affects the moisture content of various materials, influencing their properties, stability, and longevity. For example, high RH can lead to corrosion, mold growth, and the swelling of wood. Museums and storage facilities often carefully control RH to preserve artifacts.
- **Health:** Both very high and very low relative humidity can impact human health. High humidity can exacerbate respiratory issues, while low humidity can dry out skin and mucous membranes, increasing susceptibility to infections.

In summary, relative humidity provides a measure of how close the air is to being saturated with water vapor, which has significant implications for our comfort, the weather, and various natural and human-made systems.

4.2.2 Specific Humidity

Specific humidity is a measure of the mass of water vapor in a given mass of air, including the water vapor itself. It's an important concept in meteorology and climatology.

- **Formula:** Specific humidity (q) = mass of water vapor / total mass of air (including water vapor)

  q = mv / (md + mv)

  where:

  mv = mass of water vapor

  md = mass of dry air
- **Meaning:** Specific humidity represents the actual amount of water vapor present in a unit mass of air. It's usually expressed in grams per kilogram (g/kg).

4.2.3 Measurement of Relative Humidity

The relative humidity is measured using an instrument called Renault's hygrometer.

- **Description and Working**: The **Regnault's hygrometer** is a type of **dew-point hygrometer** used to measure the relative humidity of the air. It works by gradually cooling a polished metallic surface (usually silver) until dew (condensation) forms on it. The temperature at which this dew first appears is the **dew point temperature**. By comparing the dew point temperature with the ambient air temperature, the relative humidity can be determined using psychrometric tables or formulas.
- **Description:** A typical Regnault's hygrometer consists of the following components:
  - **A Polished Metallic Container (usually silver):** This container has a highly reflective surface, making it easy to observe the formation of dew.
  - **Ether:** A volatile liquid (like ether) is placed inside the metallic container.
  - **A Thermometer:** This thermometer is immersed in the ether to measure its temperature, which is assumed to be the same as the temperature of the metallic container.
  - **A Mechanism for Cooling:** Air is bubbled through the ether using a small aspirator (suction bulb). The evaporation of ether cools the container.
  - **A Second Thermometer:** This thermometer measures the ambient air temperature.
  - **A Stand:** To hold the apparatus.
  - **A Telescope:** To observe the formation of dew on the polished surface from a distance, avoiding the influence of the observer's breath.
- **Working Principle:** The working of Regnault's hygrometer is based on the principle that when air is cooled at constant pressure, the relative humidity increases. When the air cools to the point where it becomes saturated with water vapor, condensation (dew) begins to form.This temperature is the dew point. Here's how the measurement is taken:
  - **Ether is introduced into the silver container** and the thermometer inside records its temperature.
  - **Air is bubbled through the ether.**

- The rapid evaporation of the ether causes the silver container to cool down.,
- **The polished surface of the silver container is carefully observed.** As the container cools, the temperature of the air in contact with it also decreases. **The temperature at which dew just begins to form on the polished surface is noted.** This is the **dew point temperature**.
- **The bubbling of air through the ether is stopped**, and the silver container is allowed to warm up gradually.
- **The temperature at which the dew just disappears is also noted.**
- **The average of the temperature at which dew appears and disappears is taken as the dew point temperature.**
- **The ambient air temperature is recorded using the second thermometer.**
- Finally, the relative humidity is determined by using the dew point temperature and the ambient air temperature. This is typically done by referring to psychrometric tables or by using the following approximate formula:

$$RH = \frac{\text{Saturated vapor pressure at dew point}}{\text{Saturated Vapor pressure at air temperatur}} x100$$

- Saturated vapor pressure values can be found in standard tables.

4.2.4 Relevance to Meteorology and indirectly to Fisheries

While the Regnault's hygrometer directly measures atmospheric humidity, understanding atmospheric humidity is relevant to fisheries and aquaculture in several ways:

- **Evaporation from water bodies:** Humidity affects the rate of evaporation from lakes, ponds, and aquaculture systems, which in turn influences water temperature and salinity (in coastal systems).
- **Weather patterns:** Humidity is a key component of weather systems that can impact fishing activities and aquaculture infrastructure (e.g., storms associated with high humidity).
- **Disease in aquaculture:** High humidity can sometimes contribute to conditions that favor the spread of certain diseases in aquaculture facilities.

In summary, the Regnault's hygrometer is a classic instrument that provides a direct method for determining the dew point, which is then used to calculate relative humidity, an important atmospheric variable with indirect links to aquatic environments and fisheries.

4.3 Dry and Wet Bulb Thermometers

Let's detail the dry-bulb and wet-bulb thermometers used in agrometeorological observatories, their installation within a Stevenson screen, and how they are used to determine relative humidity.

4.3.1 Dry-Bulb Thermometer

- **Construction:** A dry-bulb thermometer is a standard liquid-in-glass thermometer (typically mercury or alcohol, though mercury is more common for wider temperature ranges). It consists of:
  - A glass bulb containing the liquid.
  - A glass stem with a fine bore.
  - A temperature scale etched onto the stem or on a backing attached to it.
- **Working:** The dry-bulb thermometer measures the actual temperature of the air. The bulb is exposed to the air, and the liquid inside expands or contracts with changes in air temperature. The level of the liquid in the bore corresponds to the air temperature on the scale.
- **Purpose:** It provides a direct measurement of the ambient air temperature.

4.3.2 Wet-Bulb Thermometer

- **Construction:** A wet-bulb thermometer is similar to a dry-bulb thermometer but with a crucial difference:
  - The bulb of the thermometer is covered with a muslin wick that is kept moist (usually with distilled water).
  - Like the dry-bulb thermometer, it has a glass stem with a fine bore and a temperature scale.
- **Working:** The wet-bulb thermometer measures the temperature to which air can be cooled by the evaporation of water. When air passes over the moist wick, water evaporates, which cools the bulb due to the latent heat of vaporization. The rate of evaporation, and thus the cooling, depends on the humidity of the air.
  - If the air is saturated (100% relative humidity), no net evaporation occurs, and the wet-bulb temperature will be the same as the dry-bulb temperature.

- If the air is dry, more evaporation occurs, leading to a lower wet-bulb temperature compared to the dry-bulb temperature.

- **Purpose:** In conjunction with the dry-bulb temperature, the wet-bulb temperature is used to determine the humidity of the air.

4.3.3 Installation of Dry-Bulb and Wet-Bulb Thermometers

In an agrometeorological observatory, both the dry-bulb and wet-bulb thermometers are installed inside a **Stevenson screen** to ensure they are shielded from direct solar radiation, precipitation, and ground radiation, while allowing adequate ventilation of the ambient air.

- **Mounting:** The thermometers are typically mounted vertically on a wooden frame or a specifically designed thermometer stand inside the Stevenson screen. The bulbs should be positioned at the standard height of approximately 1.25 meters above the ground.
- **Separation:** The dry-bulb and wet-bulb thermometers are placed close to each other to ensure they are measuring the temperature of the same air mass within the screen.
- **Wick Maintenance (Wet-Bulb)**
  - The muslin wick covering the bulb of the wet-bulb thermometer must be kept clean and moist with distilled water.
  - One end of the wick extends into a small reservoir of distilled water placed below the bulb. Capillary action draws water up the wick, keeping the bulb wet.
  - The water reservoir needs to be regularly checked and refilled. The wick should also be periodically cleaned or replaced to prevent contamination that could affect evaporation.

4.3.4 Method of Finding Relative Humidity

Relative humidity (RH) can be determined using the readings from the dry-bulb temperature (Td) and the wet-bulb temperature (Tw) with the help of a **psychrometric chart** or **psychrometric tables**, or by using empirical formulas.

- **Using Psychrometric Charts or Tables**
- **Read the Temperatures:** Take simultaneous readings of the dry-bulb temperature (Td) and the wet-bulb temperature (Tw).
- **Locate on the Chart/Table**

- On a psychrometric chart, find the intersection of the vertical line corresponding to the dry-bulb temperature and the diagonal line corresponding to the wet-bulb temperature. The relative humidity is then read from the curved lines on the chart passing through this intersection point.
- Psychrometric tables provide relative humidity values (and other humidity parameters) based on the dry-bulb temperature and the wet-bulb depression (Td–Tw). You would look up the row corresponding to the dry-bulb temperature and the column corresponding to the wet-bulb depression to find the relative humidity.

Water Vapor in the Atmosphere and Humidity

1. Water vapor is the gaseous form of:

A. Hydrogen B. Oxygen
C. Water D. Carbon dioxide

2. Water vapor is important in the atmosphere because it:

A. Absorbs solar radiation B. Is a greenhouse gas
C. Affects weather and climate D. All of the above

3. Which of the following holds the most water vapor?

A. Cold air B. Dry air
C. Warm air D. High-pressure air

4. The process by which water changes from liquid to vapor is:

A. Condensation B. Precipitation
C. Evaporation D. Sublimation

5. The main source of water vapor in the atmosphere is:

A. Respiration B. Oceans and lakes
C. Plants D. Volcanic eruptions

6. What happens to air's ability to hold water vapor as temperature increases?

A. Increases B. Decreases
C. Remains constant D. Becomes unstable

7. Which process contributes most to atmospheric humidity?

A. Transpiration B. Runoff
C. Infiltration D. Photosynthesis

8. The latent heat released during condensation:

A. Cools the surrounding air B. Warms the surrounding air
C. Has no effect D. Causes freezing

9. Water vapor content in the atmosphere is highest over:

A. Deserts
B. Oceans
C. Polar regions
D. Forests

10. Which is the correct range of water vapor in the atmosphere by volume?

A. 0.1% to 1%
B. 0% to 4%
C. 10% to 25%
D. 5% to 10%

11. Humidity refers to the amount of:

A. Heat in the air
B. Water vapor in the air
C. Carbon dioxide in the air
D. Oxygen in the air

12. Relative humidity is defined as:

A. Total water in the air
B. Ratio of actual to maximum vapor content
C. Water vapor in grams
D. Difference between wet and dry bulb

13. Relative humidity is expressed as:

A. Grams per cubic meter
B. Percent (%)
C. Millimeters
D. Watts per meter

14. When air is saturated, the relative humidity is:

A. 0%
B. 50%
C. 100%
D. Cannot be determined

15. Instrument used to measure humidity is called:

A. Thermometer
B. Barometer
C. Hygrometer
D. Anemometer

16. Which device uses wet and dry bulbs to measure humidity?

A. Aneroid barometer
B. Psychrometer
C. Stevenson screen
D. Pyranometer

17. If the temperature rises and moisture remains constant, relative humidity:

A. Increases
B. Decreases
C. Remains same
D. Becomes zero

18. Humidity is highest at what time of day?

A. Afternoon
B. Evening
C. Early morning
D. Midnight

19. Absolute humidity is measured in:

A. Percent
B. Grams per cubic meter
C. Kilopascal
D. Millibars

20. Which form of humidity varies most throughout the day?

A. Specific humidity
B. Absolute humidity
C. Relative humidity
D. Dew point

21. High humidity is commonly associated with:

A. Clear skies
B. Dry weather
C. Cloud formation and precipitation
D. Low temperatures only

22. Condensation occurs when:

A. Air heats up
B. Air expands
C. Dew point is reached
D. Wind speed increases

23. Fog is formed when:

A. Air temperature rises suddenly
B. Warm moist air moves over warm land
C. Cool air meets warm moist air
D. Relative humidity is very low

24. Dew forms best under:

A. Cloudy skies
B. Clear, calm nights
C. Windy conditions
D. Hot, humid days

25. High humidity makes the air feel:

A. Lighter
B. Colder
C. Warmer and sticky
D. Dry

26. When the dew point is close to air temperature:

A. Relative humidity is low
B. Fog is unlikely
C. Relative humidity is high
D. Rainfall is certain

27. Clouds are formed by:

A. Evaporation
B. Condensation of water vapor
C. Precipitation
D. Freezing

28. In weather forecasting, high humidity often signals:

A. Storms or rain
B. Cold wave
C. Windy days
D. Drought

29. Low humidity during hot weather increases:

A. Cloud cover
B. Rain chances
C. Evapotranspiration
D. Condensation

30. Which of the following contributes most to local humidity?

A. Wind
B. Water bodies
C. Soil pH
D. Dust

31. Humidity influences which of the following in crops?

A. Transpiration B. Growth and disease incidence
C. Stomatal function D. All of the above

32. High humidity in greenhouses may cause:

A. Plant dehydration B. Increased disease
C. Reduced growth D. Soil erosion

33. Low humidity in arid zones leads to:

A. Faster plant growth B. Reduced evaporation
C. Increased water loss from plants D. Cloud formation

34. Relative humidity helps in scheduling:

A. Fertilizer application B. Irrigation
C. Harvesting D. All of the above

35. For good seed storage, relative humidity should be:

A. Below 30% B. Above 70%
C. Between 50% and 70% D. Around 90%

36. High humidity encourages:

A. Fungal growth B. Soil compaction
C. Stomatal closure D. Leaf burn

37. During high humidity, transpiration in plants is:

A. Increased B. Decreased
C. Unaffected D. Stops completely

38. Humidity influences the drying of:

A. Leaves B. Soil
C. Grains and fodder D. Insects

39. Evapotranspiration is the sum of:

A. Condensation + evaporation B. Runoff + infiltration
C. Evaporation + transpiration D. Transpiration + absorption

40. High humidity can delay:

A. Flowering B. Germination
C. Maturity D. Harvesting and drying

41. Humidity contributes to climate by:

A. Regulating rainfall B. Modifying temperatures
C. Controlling cloud formation D. All of the above

42. Human discomfort increases with:

A. High humidity and high temperature
B. High wind and low humidity
C. Cold and dry weather
D. Low temperature and low humidity

43. The Heat Index combines:

A. Temperature and pressure
B. Humidity and temperature
C. Wind and radiation
D. Air density and CO_2

44. Humidity affects animal comfort through:

A. Body heat regulation
B. Sweating and evaporation
C. Respiratory efficiency
D. All of the above

45. Humidity control is important in:

A. Cold storage
B. Poultry farms
C. Food processing
D. All of the above

46. Air conditioning systems lower humidity by:

A. Adding water vapor
B. Heating the air
C. Cooling and condensing moisture
D. Using oxygen

47. Which factor lowers indoor humidity naturally?

A. Fans
B. Dehumidifiers
C. Open windows
D. Both B and C

48. Comfortable relative humidity range for humans indoors is:

A. 10–20%
B. 30–60%
C. 60–90%
D. Above 90%

49. Excessive humidity indoors can cause:

A. Dehydration
B. Mold and mildew
C. Low oxygen
D. Fire risk

50. Water vapor contributes to the greenhouse effect by:

A. Blocking visible light
B. Trapping longwave radiation
C. Absorbing UV rays
D. Reflecting heat back to space

Answer Key (s)

1	c	2	d	3	c	4	c	5	b	6	a	7	a	8	b	9	b	10	B
11	b	12	b	13	b	14	c	15	c	16	b	17	b	18	c	19	b	20	c
21	c	22	c	23	c	24	b	25	c	26	c	27	b	28	a	29	c	30	b
31	d	32	b	33	c	34	d	35	a	36	a	37	b	38	c	39	c	40	d
41	d	42	d	43	b	44	d	45	d	46	c	47	d	48	b	49	b	50	b

5

Cloud Formation and Classification

Clouds, those ever-present features of our sky, are more than just aesthetically pleasing; they are fundamental components of Earth's weather and climate systems. This chapter, "Cloud Formation and Classification," will embark on a journey to understand the fascinating processes that lead to their creation and the systematic ways in which they are categorized.

The essential ingredients for cloud formation are the presence of water vapor in the atmosphere, the availability of condensation or ice nuclei, and the crucial mechanism of air cooling. We will look into how rising air parcels cool adiabatically, reaching saturation and allowing water vapor to condense or deposit into the myriad forms we recognize as clouds.

Following the understanding of their genesis, we will navigate the internationally recognized system of cloud classification. This system, primarily based on a cloud's altitude, appearance, and the processes that formed it, provides a common language for meteorologists and observers worldwide. We will learn to identify the main cloud types – high, mid-level, low, and vertically developing – and their associated characteristics, from the wispy cirrus to the towering cumulonimbus.

Furthermore, we will touch upon the significance of different cloud types in indicating current weather conditions and forecasting future weather. The appearance and evolution of clouds can offer valuable clues about atmospheric stability, moisture content, and the likelihood of precipitation.

In the end of the chapter, you will gain a comprehensive understanding of how clouds form, the diverse array of cloud types that grace our skies, and their importance in the broader context of atmospheric science. Let's unlock the secrets held within these visible masses of water and ice suspended in our atmosphere.

5.1 Clouds

Clouds are visible masses of tiny water droplets or ice crystals suspended in the Earth's atmosphere. These droplets or crystals are so small and light that they can remain aloft.What we see as a cloud is essentially a large collection of these minute particles.

5.1.1 Formation of Couds

Clouds form when invisible water vapor in the air condenses into liquid water droplets or freezes into ice crystals. This happens when air becomes saturated, meaning it can no longer hold all of its water vapor (Fig.5.1). Saturation is typically reached in one of two ways:

- **Cooling the Air:** As air rises in the atmosphere, it expands due to lower air pressure at higher altitudes. This expansion causes the air to cool. Cooler air can hold less water vapor than warmer air. When the air cools to its **dew point** (the temperature at which the air becomes saturated), the excess water vapor condenses.
- **Adding Moisture to the Air:** If more water vapor is added to the air (e.g., through evaporation from a body of water), the air can eventually become saturated at its current temperature, leading to condensation.

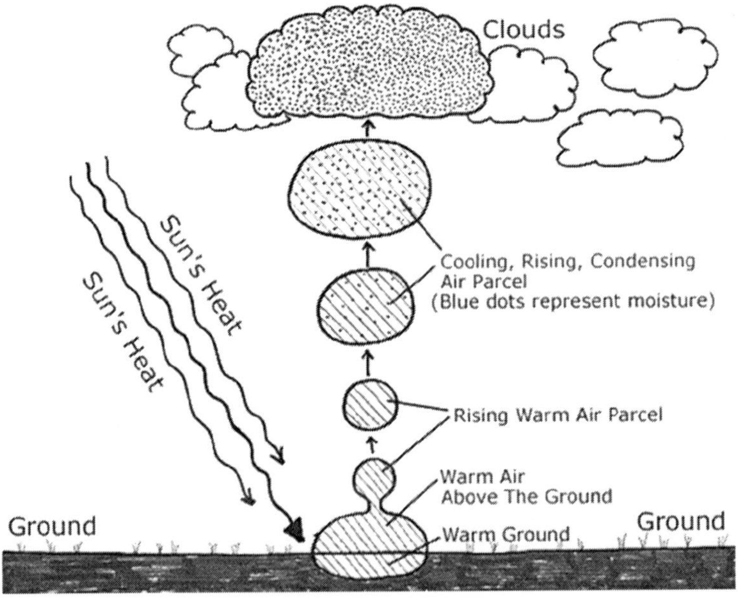

Fig. 5.1: Formation of clouds

For condensation (or deposition, the direct formation of ice from vapor) to occur, water vapor needs a surface to condense onto. These microscopic airborne particles, such as dust, pollen, salt crystals, and pollutants, are called **condensation nuclei** (or ice nuclei if the temperature is below freezing). Water vapor molecules cluster around these nuclei to form the tiny cloud droplets or ice crystals.

5.1.2 Growth of Clouds

Once the initial cloud droplets or ice crystals form, clouds can grow through several processes:

- **Further Condensation/Deposition:** If the air continues to cool or more moisture is added, more water vapor will condense or deposit onto the existing droplets or crystals, causing them to grow larger.
- **Collision and Coalescence:** Within a cloud, droplets of different sizes fall at slightly different speeds. This can lead to collisions between droplets. When they collide, they can coalesce (merge) into larger droplets. This process is particularly important in warmer clouds for the formation of raindrops.
- **Bergeron-Findeisen Process (Ice-Crystal Process):** In colder clouds (where both ice crystals and supercooled water droplets coexist), ice crystals tend to grow at the expense of water droplets. This is because the saturation vapor pressure over ice is lower than that over liquid water at the same temperature. Water vapor will deposit onto the ice crystals more readily than it will condense onto the water droplets, causing the ice crystals to grow. These ice crystals can then fall and melt into rain, or remain as snow or other forms of frozen precipitation.
- **Uplift:** Continued lifting of air can lead to more condensation as the air cools at higher altitudes, adding more moisture to the cloud and increasing its vertical extent. This is particularly important for the development of towering clouds like cumulonimbus.

The growth of clouds is a dynamic process involving the interplay of thermodynamics, microphysics (the behavior of tiny water and ice particles), and atmospheric motion.

5.1.3 Dissipation of Clouds

Dissipation of clouds refers to the process by which clouds weaken, shrink, and eventually disappear from the sky. This happens when the conditions that led to their formation and growth cease to be effective, and the balance shifts towards the evaporation or sublimation of the water droplets or ice crystals within them. Here are the main ways clouds dissipate:

- **Warming of the Air:** If the temperature of the air within or around the cloud increases, the air's capacity to hold water vapor rises. This can lead to the evaporation of the water droplets or sublimation of the ice crystals in the cloud, causing it to shrink and disappear. Solar heating, especially for low clouds, or the descent of air (which warms adiabatically) can cause this.

- **Mixing with Drier Air (Entrainment):** Clouds don't exist in isolation; they constantly interact with the surrounding air. If the air surrounding a cloud is significantly drier (lower relative humidity), the water droplets in the cloud will evaporate into this drier air in an attempt to reach equilibrium. This mixing of dry air into the cloud, known as entrainment, can erode the cloud over time, making its edges appear wispy and less defined until it vanishes.
- **Sinking Air (Subsidence):** When air sinks in the atmosphere, it compresses and warms adiabatically. This warming increases the air's capacity to hold water vapor, leading to the evaporation of cloud droplets. Large-scale sinking motion associated with high-pressure systems can cause widespread cloud dissipation.
- **Precipitation:** If a cloud produces rain, snow, or other forms of precipitation, it loses some of its water content. If the rate of precipitation exceeds the rate at which water vapor is condensing within the cloud, the cloud will eventually dissipate as its water supply is depleted. Essentially, the cloud "rains itself out."

Often, the dissipation of a cloud involves a combination of these processes working simultaneously. For example, a low cloud might dissipate due to a combination of daytime heating and mixing with slightly drier air above it.

5.1.4 Classification of Clouds

The primary criteria used for the classification of clouds are their **altitude of formation** in the atmosphere and their **appearance or physical form**. This system, largely based on the work of Luke Howard in the early 19th century and refined by the World Meteorological Organization (WMO), allows for a systematic understanding and identification of different cloud types. Here's a breakdown of the criteria (Fig.5.2):

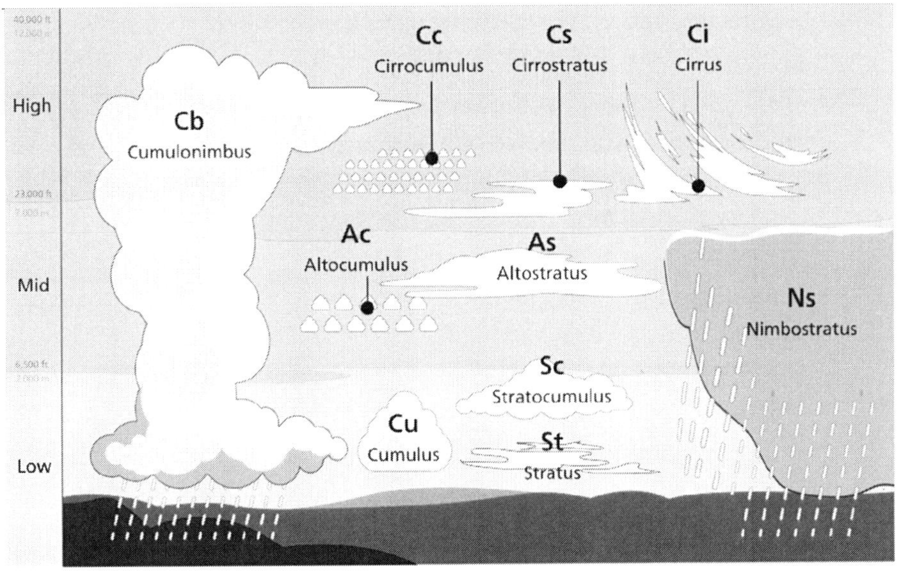

Fig. 5.2: Clouds classification

Altitude of Formation: Clouds are broadly categorized into three main height levels (though some extend through multiple levels).

- **High-Level Clouds:** Typically form above 5,000 meters (16,500 feet) in temperate regions. At these altitudes, the air is very cold, so these clouds are primarily composed of ice crystals. They are given the prefix "cirro-". Examples include Cirrus (Ci), Cirrocumulus (Cc), and Cirrostratus (Cs).
- **Mid-Level Clouds:** Form between 2,000 and 7,000 meters (6,500 to 23,000 feet) in temperate regions. These clouds are composed of water droplets, ice crystals, or a mixture of both (often supercooled water). They are given the prefix "alto-". Examples include Altocumulus (Ac) and Altostratus (As).
- **Low-Level Clouds:** Form from the surface up to 2,000 meters (6,500 feet). These are mainly composed of water droplets, but can contain ice crystals in colder regions or during winter. Examples include Stratus (St), Stratocumulus (Sc), and Nimbostratus (Ns - which also has vertical development).
- **Clouds with Vertical Development:** These clouds don't fit neatly into a single height category as their bases are typically in the low to mid-levels, but their tops can extend high into the atmosphere. They are associated with instability in the atmosphere. Examples include Cumulus (Cu) and Cumulonimbus (Cb).

The altitude ranges can vary slightly depending on latitude (they tend to be higher in the tropics and lower in polar regions).

- **Appearance or Physical Form:** The second key criterion is the visual characteristics of the clouds, which describe their shape, texture, and opacity. Several root names are used to describe these forms.
  - **Cirrus (Ci):** Thin, wispy, and feathery clouds composed of ice crystals. The term "cirro-" is used for high-level clouds of this nature.
  - **Cumulus (Cu):** Puffy, cotton-like clouds with flat bases and dome-shaped tops, indicating rising air. The term "cumulo-" is used in combination with other height indicators.
  - **Stratus (St):** Flat, hazy, and featureless sheet-like clouds that often cover the entire sky. The term "strato-" is used for low-level clouds of this nature.
  - **Nimbus (Nimbo-):** Indicates a cloud that is producing or capable of producing precipitation (rain or snow). This prefix or suffix is added to other cloud type names (e.g., Nimbostratus, Cumulonimbus).

By combining these altitude-based prefixes/names with the form-based names, we get the names of the ten basic cloud genera recognized internationally. For example:

- **Cirrostratus:** High-level (cirro-) and sheet-like (stratus).
- **Altocumulus:** Mid-level (alto-) and puffy (cumulus).
- **Nimbostratus:** Low to mid-level (stratus-based) and rain-bearing (nimbo-).
- **Cumulonimbus:** Vertically developing (cumulo-) and rain-bearing (nimbus).

Beyond these genera, clouds are further classified into **species** (describing more specific shapes and internal structures) and **varieties** (describing the arrangement and transparency). Additional features like precipitation (virga, praecipitatio), and accessory clouds (pileus, mamma) are also noted.

In summary, the classification of clouds primarily relies on **how high they are** and **what they look like**, providing a structured way to understand and communicate about these important atmospheric phenomena.

5.1.5 Description of Main Genera of Clouds

Let us rewrite the descriptions for the main cloud genera, incorporating both their altitude and appearance:

- **High-Level Clouds (composed primarily of ice crystals)**
  - **Cirrus (Ci):** These are delicate, white, often fibrous or silky clouds with a wispy appearance. They often look like streaks or thin, detached patches with hair-like filaments. Being high up, they are made of ice crystals and typically appear against a blue sky. They can indicate an approaching warm front.
  - **Cirrocumulus (Cc):** Appearing high in the atmosphere, these clouds look like small, white patches arranged in ripples or waves, sometimes described as a "mackerel sky." They are composed of ice crystals and can sometimes show iridescence (colors).
  - **Cirrostratus (Cs):** These are thin, sheet-like, high-level clouds composed of ice crystals. They often appear as a pale, milky veil that can cover the entire sky. A common characteristic is the halo effect they can produce around the sun or moon due to the refraction of light through their ice crystals.
- **Mid-Level Clouds (composed of water droplets, ice crystals, or a mixture)**
  - **Altocumulus (Ac):** These mid-level clouds appear as white or gray patches, often arranged in layers or waves. They can have a puffy or somewhat fibrous appearance. Sometimes, parts of the cloud may be darker than others. They are often described as being larger and thicker than cirrocumulus.
  - **Altostratus (As):** These are gray or bluish-gray mid-level sheets that often cover the entire sky. The sun or moon may be visible as a dim, watery disk without a halo. Altostratus clouds are typically more uniform and less structured than altocumulus. They can indicate an approaching storm system.
- **Low-Level Clouds (composed primarily of water droplets)**
  - **Stratus (St):** These are gray, uniform, sheet-like low-level clouds that can often cover the entire sky, resembling a featureless overcast. They may produce light drizzle or mist. When stratus clouds are very low and in contact with the ground, they are called fog.
  - **Stratocumulus (Sc):** These low-level clouds appear as patches or layers of rounded masses, rolls, or ripples. They are generally gray or whitish with darker parts. Stratocumulus clouds are more defined and have more texture than stratus clouds. They rarely produce significant precipitation.

- **Nimbostratus (Ns):** These are dark, gray, featureless rain clouds (or snow clouds). They are typically thick enough to block out the sun entirely and are associated with continuous, light to moderate precipitation that can last for several hours. While their bases are low, they can also extend into the mid-levels.

• Clouds with Vertical Development (bases in low to mid-levels, tops can extend high)

- **Cumulus (Cu):** These are puffy, dome-shaped clouds with flat bases. They are often described as looking like cotton balls. Fair-weather cumulus clouds (Cumulus humilis) are small and don't typically produce precipitation. However, they can grow vertically into larger Cumulus congestus, which can bring showers.
- **Cumulonimbus (Cb):** These are large, towering, vertically developed clouds associated with thunderstorms. Their tops can reach very high into the atmosphere, often taking on an anvil shape (incus). Cumulonimbus clouds are capable of producing heavy rain, lightning, thunder, hail, and even tornadoes. Their bases are often dark and menacing.

Cloud Formation and Classification

1. What is the primary factor responsible for cloud formation?

a) Wind speed b) Solar radiation

c) Condensation of water vapor d) Earth's rotation

2. Which process causes water vapor to become liquid water in the atmosphere?

a) Evaporation b) Sublimation

c) Condensation d) Precipitation

3. Clouds form when air becomes saturated and

a) Loses oxygen b) Rises and cools

c) Becomes denser d) Falls and heats up

4. What is the lifting condensation level (LCL)?

a) Height where dew forms b) Altitude where rising air becomes saturated

c) Point of maximum cloud thickness d) Level where rainfall starts

5. Which of the following is a high-level cloud?

a) Stratus b) Nimbostratus

c) Cirrus d) Cumulus

6. Which clouds are usually associated with thunderstorms?

a) Altostratus b) Cirrostratus

c) Cumulonimbus d) Stratocumulus

7. **Cirrostratus clouds are composed primarily of**
   a) Ice crystals b) Dust particles
   c) Liquid water d) Water vapor
8. **What type of cloud has a puffy appearance and flat base?**
   a) Stratus b) Cumulus
   c) Nimbus d) Cirrus
9. **Which type of cloud covers the sky uniformly like a blanket?**
   a) Stratus b) Cumulus
   c) Cirrus d) Altocumulus
10. **Altostratus clouds are found at which altitude range?**
    a) Below 2,000 m b) 2,000 to 7,000 m
    c) 7,000 to 13,000 m d) Above 13,000 m
11. **Which cloud type produces continuous rain or snow?**
    a) Cirrocumulus b) Nimbostratus
    c) Altocumulus d) Stratocumulus
12. **Clouds that resemble fish scales are**
    a) Cirrocumulus b) Altostratus
    c) Nimbostratus d) Cumulus
13. **Which cloud type indicates fair weather but may develop into storm clouds?**
    a) Cumulus b) Stratus
    c) Cirrostratus d) Altostratus
14. **The vertical development of cumulonimbus clouds is due to**
    a) Strong convection b) Temperature inversion
    c) Low wind speed d) Stable atmosphere
15. **The prefix "cirro-" refers to clouds at**
    a) Low altitude b) Medium altitude
    c) High altitude d) Any level
16. **Stratocumulus clouds appear as**
    a) Low, lumpy layers b) High wispy clouds
    c) Towering thunderheads d) Thin sheets of ice
17. **What does "nimbus" indicate in cloud classification?**
    a) Puffy shape b) High altitude
    c) Precipitation d) Warm air
18. **Which cloud type forms due to gentle uplift over a large area?**
    a) Stratus b) Cumulonimbus
    c) Cirrocumulus d) Altocumulus

19. Clouds form best in air that is

a) Stable and dry
b) Moist and unstable
c) Warm and dry
d) Cool and stable

20. The process of orographic lifting involves

a) Heating from below
b) Air rising over a mountain
c) Frontal convergence
d) Sea breeze

21. Which cloud type is often associated with halos around the sun or moon?

a) Stratocumulus
b) Cirrostratus
c) Nimbostratus
d) Cumulus

22. What type of cloud usually signals an approaching warm front?

a) Altostratus
b) Cumulus
c) Stratocumulus
d) Cirrocumulus

23. The presence of cumulonimbus clouds indicates

a) Dry weather
b) Severe weather
c) Fog
d) Calm conditions

24. Which cloud is associated with fog at ground level?

a) Stratus
b) Cirrus
c) Altocumulus
d) Cumulus

25. Cloud droplets form around small particles called

a) Ice crystals
b) Condensation nuclei
c) Water vapor
d) Oxygen molecules

26. The altitude of cloud base in cumulus clouds is determined by

a) Dew point and temperature difference
b) Solar radiation
c) Latitude
d) Humidity alone

27. Cloud classification was first formally proposed by

a) Luke Howard
b) Aristotle
c) Galileo
d) Halley

28. Cumulonimbus clouds can reach altitudes up to

a) 3,000 m
b) 6,000 m
c) 12,000 m or more
d) 2,000 m

29. Lenticular clouds form due to

a) Airflow over mountains
b) Sea breeze
c) Convective heating
d) Frontal lifting

30. A fog is technically a

a) Stratus cloud near ground
b) Nimbostratus cloud
c) Low-level cirrostratus
d) Cumulus cloud base

31. Clouds are primarily made up of

a) Tiny water droplets and/or ice crystals
b) Oxygen molecules
c) Dust and smoke
d) Nitrogen gas

32. Clouds appear white because

a) They reflect all visible light wavelengths
b) They absorb sunlight
c) They emit UV light
d) They scatter only red light

33. Dark clouds indicate

a) They are far away
b) Thick clouds with less light penetration
c) Presence of carbon particles
d) High-speed wind

34. Which of the following does *not* influence cloud formation?

a) Temperature
b) Magnetic field
c) Humidity
d) Air pressure

35. Cloud classification is primarily based on

a) Shape and altitude
b) Temperature and color
c) Size and speed
d) Location and density

36. In aviation, cumulonimbus clouds are considered

a) Hazardous due to turbulence
b) Useful for lift
c) Calm and safe
d) Indicators of clear weather

37. The cloud base is generally lower in

a) Humid climates
b) Dry regions
c) Mountainous terrain
d) Winter

38. Which cloud appears as a thin veil covering the sky?

a) Cirrostratus
b) Cumulus
c) Nimbostratus
d) Altocumulus

39. Which type of cloud is often observed after a cold front passes?

a) Cumulus
b) Cirrostratus
c) Altostratus
d) Nimbostratus

40. The term "cloud deck" refers to

a) A continuous layer of clouds
b) Clouds over the sea
c) High cumulonimbus
d) Scattered cloud formations

41. Cloud albedo refers to

a) Reflectivity of clouds
b) Heat trapping by clouds
c) Wind speed near clouds
d) Water content in clouds

42. Which cloud is typically highest in the troposphere?

a) Cirrus
b) Stratus
c) Cumulus
d) Nimbostratus

43. Low-level clouds are found at altitudes

a) Below 2,000 meters
b) 2,000 to 6,000 meters
c) 6,000 to 13,000 meters
d) Above 13,000 meters

44. Which process leads to cloud dissipation?

a) Evaporation
b) Precipitation
c) Accumulation
d) Sublimation

45. Which cloud type has the greatest vertical extent?

a) Stratus
b) Cirrus
c) Cumulonimbus
d) Altostratus

46. How are clouds important for Earth's energy balance?

a) They reflect solar radiation and trap heat
b) They generate heat
c) They absorb wind energy
d) They increase evaporation

47. Which instrument is used to observe cloud height?

a) Ceilometer
b) Barometer
c) Anemometer
d) Hygrometer

48. Clouds are least likely to form in

a) Very dry and stable air
b) Moist and unstable air
c) Rising air masses
d) Coastal areas

49. Which cloud often forms in advance of a storm system?

a) Altostratus
b) Cumulus
c) Stratocumulus
d) Cirrocumulus

50. In satellite images, thick cloud layers appear

a) Bright white
b) Blue
c) Transparent
d) Black

Answer Key (s)

1	c	2	c	3	b	4	b	5	c	6	c	7	a	8	b	9	a	10	b
11	b	12	a	13	a	14	a	15	c	16	a	17	c	18	a	19	b	20	b
21	b	22	a	23	b	24	a	25	b	26	a	27	a	28	c	29	a	30	a
31	a	32	a	33	b	34	b	35	a	36	a	37	a	38	a	39	a	40	a
41	a	42	a	43	a	44	a	45	c	46	a	47	a	48	a	49	a	50	a

6

Precipitation and its Forms

Precipitation is a vital component of the Earth's water cycle, playing a crucial role in shaping our climate, weather, and ecosystems. It occurs when water vapor in the atmosphere condenses and falls to the ground as rain, snow, sleet, or hail. Precipitation is a key driver of weather patterns, influencing everything from local weather events to global climate trends.

In this chapter, we will explore the various forms of precipitation, including rain, snow, sleet, and hail, and examine the factors that influence their formation and distribution. We will also discuss the importance of precipitation in sustaining life on Earth, from supporting plant growth and agriculture to shaping landscapes and ecosystems. The key topics are:

- Forms of precipitation (rain, snow, sleet, hail)
- Precipitation processes (formation, distribution, intensity)
- Factors influencing precipitation (temperature, humidity, wind patterns)
- Importance of precipitation (ecosystems, agriculture, water resources)
- Precipitation measurement and prediction (methods, challenges, applications)

By understanding precipitation, we can better appreciate the complex interactions between the atmosphere, oceans, and land surfaces that shape our planet's climate and weather patterns.

6.1 Forms of Precipitation

Let's break down the different forms of precipitation (Fig.6.1):

6.1.1 Rain

Rain is a type of precipitation that occurs when water droplets in clouds grow and become too heavy to remain suspended in the air. Rain droplets are typically larger than 0.5 mm in diameter. Rainfall can be classified into different types, including convectional, orographic, and cyclonic rainfall.

6.1.2 Drizzle

Drizzle is a type of light rainfall with drop sizes less than 0.5 mm. It's often steady and continuous, but with much smaller droplets than regular rain.

6.1.3 Snow

Snow forms when water vapor in clouds freezes into ice crystals. Snowfall occurs when the temperature is below 0°C, and the ice crystals stick together to form snowflakes. Snow can be heavy or light, and its intensity can vary greatly.

6.1.4 Sleet

Sleet forms when snowflakes fall through a layer of warm air, causing them to melt into raindrops. If these raindrops then pass through a layer of cold air before reaching the ground, they freeze into small, transparent ice pellets called sleet.

6.1.5 Hail

Hail forms when updrafts in thunderstorms carry water droplets up into the freezing level of the atmosphere, where they freeze into small balls of ice. As the hailstones move upward and downward through the storm cloud, they may pass through additional layers of water droplets and freezing air, causing them to grow in size.

Fig. 6.1: Forms of precipitation

6.1.6 Frost

Frost forms when the air temperature cools to its dew point, causing the water vapor to condense and freeze onto surfaces. This can occur when the air temperature is below freezing, resulting in a layer of ice crystals on surfaces.

6.1.7 Dew

Dew forms when the air temperature cools to its dew point, causing the water vapor to condense onto surfaces. This typically occurs overnight as the ground rapidly cools, causing the air near the surface to cool and reach its dew point.

6.1.8 Other Forms of Precipitation

- **Virage**: Raindrops that evaporate before reaching the ground, often occurring when rain falls through dry air.
- **Mist**: A type of fog that forms when evaporation occurs before reaching the ground, reducing visibility.
- **Hailstones**: Large balls of ice that form in thunderstorms and can cause significant damage.

• Precipitation forms through various processes, including

- **Convectional Precipitation**: Heated air rises, cools, and condenses, forming clouds and precipitation.
- **Orographic Precipitation**: Moist air is forced to rise over mountains, cooling and condensing to form clouds and precipitation.
- **Cyclonic Precipitation**: Large-scale precipitation that occurs when air masses with different temperatures and humidity levels meet.

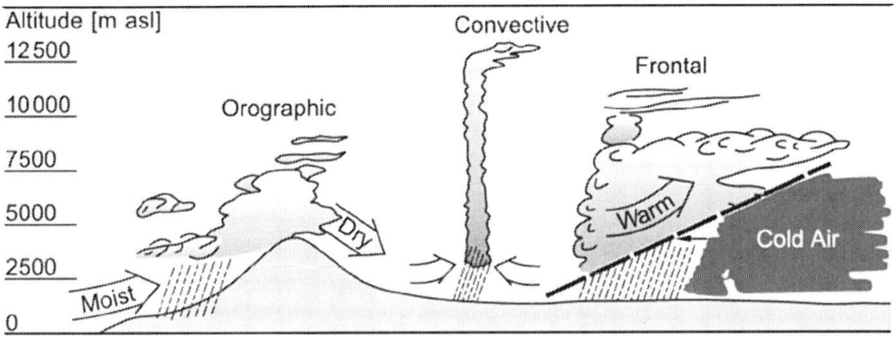

Fig. 6.2: Precipitation through various processes

6.2 Convectional Precipitation

Convectional precipitation is a type of precipitation that occurs when the sun's energy heats the surface of the Earth, causing water to evaporate and form water vapor. This process leads to the formation of clouds and precipitation through a series of stages.

6.2.1 The Process of Convectional Precipitation

- **Evaporation:** The sun heats the ground, causing water to evaporate from surfaces such as oceans, lakes, and rivers. This water vapor rises into the air as gas.

- **Rising Air:** As the ground continues to heat up, it warms the air above it, causing it to expand and rise in convection currents. This warm air is less dense than the surrounding air, allowing it to rise.
- **Cooling and Condensation:** As the air rises, it cools, and the water vapor in the air condenses into tiny water droplets, forming clouds.
- **Cloud Formation:** The condensed water droplets gather together to form visible clouds. Cumulonimbus clouds are a common type of cloud associated with convectional precipitation.
- **Precipitation:** When the water droplets in the clouds become too heavy to remain suspended in the air, they fall to the ground as precipitation, often in the form of heavy rain or thunderstorms.

6.2.2 Characteristics of Convectional Precipitation

- **Heavy Rainfall:** Convectional precipitation is characterized by heavy rainfall, often accompanied by thunder and lightning.
- **Short-Lived:** This type of precipitation is typically short-lived, lasting from a few minutes to a few hours.
- **Localized:** Convectional precipitation is highly localized, often occurring in specific areas where the ground is heated by the sun.
- **Associated with Cumulonimbus Clouds:** Convectional precipitation is often associated with cumulonimbus clouds, which can produce heavy precipitation and thunderstorms.

6.2.3 Regions Prone to Convectional Precipitation

- **Tropical Regions:** Convectional precipitation is common in tropical regions near the equator, where the sun's energy is strongest.
- **Equatorial Regions:** Regions like the Amazon Rainforest, Congo Basin, and islands of Southeast Asia experience convectional precipitation due to their location near the equator.
- **Temperate Regions During Summer:** Areas like South East England can experience convectional precipitation during warm sunny spells in the summer.

6.2.4 Importance of Convectional Precipitation

- **Water Cycle:** Convectional precipitation plays a crucial role in the water cycle, helping to distribute water around the globe.
- **Agriculture:** This type of precipitation can provide much-needed water for crops, especially in regions with limited rainfall.

- **Weather Patterns:** Convectional precipitation can influence local weather patterns, leading to the formation of thunderstorms and heavy rainfall.

In conclusion, convectional precipitation is an important process that helps shape our climate and weather patterns. Understanding this process can provide valuable insights into the complex interactions between the atmosphere, oceans, and land surfaces.

6.3 Orographic Precipitation

Orographic precipitation occurs when a saturated air mass encounters a mountain or orographic barrier, forcing it to rise and cool, resulting in condensation and precipitation. This type of precipitation is commonly observed in mountainous regions where moist air is forced to ascend, leading to cloud formation and precipitation.

6.3.1 The Process of Orographic Precipitation

- **Air Mass Encounter**: A saturated air mass encounters a mountain or orographic barrier, such as a mountain range.
- **Forced Ascent**: The air mass is forced to rise due to the initial momentum, leading to a decrease in pressure and temperature.
- **Condensation**: As the air mass gains height, the water vapor condenses into clouds, and precipitation occurs when the clouds become saturated.
- **Precipitation Distribution**: The windward slopes of the mountain receive more precipitation, while the leeward slopes remain dry due to the descending air.

6.3.2 Characteristics of Orographic Precipitation

- **Windward Slopes Receive More Rainfall**: Mountains and hills can force warm, moist air to rise, cool, and condense, resulting in precipitation on the windward side.
- **Rain Shadow Effect**: The leeward side of the mountain receives less rainfall due to the descending air, creating arid or semi-arid regions.
- **Localized Precipitation**: Orographic precipitation is highly localized, often occurring in specific areas where the terrain forces air to rise.

6.3.3 Examples of Orographic Precipitation

- **Mahabaleshwar, India**: Situated on the Western Ghats, receives over 600 cm of rainfall annually, while Pune, in the rain shadow area, receives only about 70 cm.

- **Patagonian Desert, Argentina**: An example of a region affected by the rain shadow effect, resulting in arid conditions.

6.3.4 Importance of Orographic Precipitation

- **Water Cycle**: Orographic precipitation plays a crucial role in the water cycle, distributing water around the globe.
- **Ecosystems**: This type of precipitation supports plant growth and agriculture in regions with limited rainfall.
- **Climate Regulation**: Orographic precipitation influences local climate patterns, shaping regional weather conditions .

6.4 Cyclonic Precipitation

The process of cyclonic precipitation involves several steps:

- **Low Pressure Formation**: A low-pressure system develops due to the warming of air near the surface or the convergence of air masses.
- **Air Rise**: As air rises in the low-pressure system, it moves into cooler parts of the atmosphere.
- **Condensation**: The rising air cools, and the water vapor condenses into clouds.
- **Precipitation**: When the clouds are saturated with water, precipitation occurs.

6.4.1 Characteristics

- **Intensity and Duration**: Cyclonic precipitation can vary widely in intensity and duration. Tropical cyclones are known for intense, short-duration rainfall, while temperate cyclones can bring prolonged periods of rain or snow.
- **Distribution**: The precipitation is often widespread around the center of the cyclone but can be more concentrated near the eye wall in tropical cyclones.
- **Impact**: Cyclonic precipitation can lead to significant flooding, damage from strong winds, and disruptions in transportation and daily life.

6.5 Measurement of Precipitation

The rainfall is measured using a rain gauge.

6.5.1 Rain gauge

The India Meteorological Department (IMD) primarily uses non-recording and recording type rain gauges for rainfall measurement. Here's a detailed look at the construction, installation, and measurement procedures, mainly focusing on the commonly used non-recording Symon's type and the automatic Tipping Bucket Rain Gauge (TBRG).

- **Construction**
  - **Non-Recording Rain Gauge (Symon's Type / FRP Rain Gauge):** IMD has shifted to using Fiber-glass Reinforced Polyester (FRP) rain gauges, which come in two variants with collecting areas of 100cm2 and 200cm2. However, the principle remains similar to the Symon's gauge. A standard non-recording rain gauge generally consists of three parts:
    - **Base:** A splayed base that is fixed onto a foundation.
    - **Collector (Funnel):** A circular funnel with a sharp-edged rim (127 mm diameter for Symon's type, or dimensions corresponding to 100cm2 or 200cm2 area for FRP type). The funnel directs the rainwater into the receiver. The slope of the funnel is steep (at least 45°) to prevent rain from splashing out.
    - **Receiver (Bottle/Cylinder):** A container placed inside the body of the gauge to collect the rainwater that passes through the funnel. Some gauges have an additional cylinder to prevent overflow. The receiver often has a narrow neck to minimize evaporation losses.

The base and the collector are locked together. Modern FRP gauges often include a hasp and staple for permanent locking with a padlock to prevent tampering.

- **Recording Rain Gauge (Tipping Bucket Rain Gauge - TBRG)**

The TBRG used by IMD typically consists of:

- **Collector:** A funnel similar to the non-recording type, which collects rainwater. It usually has a larger orifice (e.g., 200 mm diameter).
- **Tipping Bucket Mechanism:** Below the funnel, there is a two-compartment bucket pivoted such that when one compartment fills with a specific amount of water (e.g., 0.5 mm of rainfall equivalent), it tips, emptying the water and bringing the other compartment under the funnel.
- **Recording System:** Each tip of the bucket closes an electrical circuit momentarily, producing a signal (a pulse). This signal is recorded by a data logger, which counts the number of tips over a period. The data logger stores this information, often with a timestamp, allowing for the determination of rainfall intensity and duration.

- **Housing:** An enclosure to protect the mechanism and electronics from the weather.

- **Installation**

As per IS 4986:2002 (which aligns with IMD practices), the installation of a non-recording rain gauge involves the following:

- **Foundation:** The rain gauge should be fixed on a masonry or concrete foundation of size 600 mm × 600 mm × 600 mm sunk into the ground. Steel structures can also be used.
- **Leveling and Height:** The base of the gauge is cemented into this foundation such that the rim of the collector is exactly 300 mm (30 cm) above the ground level and perfectly horizontal. This height minimizes splashing and wind effects.
- **Exposure**
  - The gauge should be installed on level ground, away from slopes or terraces, and never on a wall or roof (unless specific conditions necessitate and precautions are taken).
  - The distance between the rain gauge and any nearby object (like a tree or building) should generally be at least twice the height of the object, preferably four times the height.
  - In mountainous or coastal regions, care should be taken to shield the gauge from strong winds using a belt of trees or a wall at a suitable distance.
- **Protection:** A fence should be erected around the rain gauge to protect it from damage by stray animals and miscreants. The height of the fence should not be more than half the distance of the fence from the gauge.
- **Locking and Maintenance:** The rain gauge should be kept locked and periodically painted to prevent corrosion.

- **Height**

- The standard height for the rim of the collector above the ground is usually around 30 cm (though this can vary slightly depending on local practices). This height is a compromise to minimize splash from the ground into the gauge while also minimizing wind effects that might affect the catch.

- **Secure Mounting**
  - The rain gauge should be securely mounted on a stable base or fixed to the ground to prevent it from being knocked over by wind or animals.

6.6 Measurement of Rainfall

Rainfall is typically measured once every 24 hours, usually at a fixed observation time (e.g., 08:30 IST in India). The process involves:

- **Collection:** The rainwater that falls into the collector is funneled into the receiving cylinder.
- **Transfer to Measuring Cylinder:** Carefully lift the receiving cylinder and pour the collected rainwater into the graduated measuring cylinder. To avoid spillage, pour slowly and steadily.
- **Reading the Measurement**
  - Place the measuring cylinder on a level surface.
  - Read the level of the water in the measuring cylinder at the bottom of the meniscus (the curved upper surface of the water).
  - The graduations on the measuring cylinder will directly give the amount of rainfall in millimeters (mm). For example, if the water level reaches the mark of 15.5 mm, the rainfall for that period is recorded as 15.5 mm.
  - If the amount of rainfall exceeds the capacity of the measuring cylinder, empty it and continue measuring the remaining water in the receiving cylinder, adding the readings together.
- **Recording:** Record the measured rainfall amount along with the date and time of observation in the meteorological register. If no rain has fallen, it should be recorded as 0.0 mm.
- **Emptying:** After measurement, the receiving cylinder should be emptied and placed back under the funnel, ready for the next rainfall event.

This type of non-recording rain gauge provides the total amount of rainfall over a period but does not give information about the intensity or the time distribution of the rainfall. For such information, recording-type rain gauges are used. In summary, IMD specifications emphasize proper site selection, careful installation to ensure accurate collection, and standardized procedures for measuring and recording rainfall using both non-recording and recording instruments.

6.6.1 Automatic Rainfall Recorder

Let's describe a **Float-Type (Natural Syphon) Automatic Rainfall Recorder**, as it directly uses the rise in water level for recording, which aligns with the term "hydrographic" (related to water).

- **Construction**
  - **Collector:** A funnel at the top with a sharp-edged rim of a known area (e.g., 200 cm^2) to collect rainwater.
  - **Float Chamber:** A cylindrical chamber below the funnel where the collected rainwater accumulates.
  - **Float:** A light, hollow float that rests on the surface of the water in the float chamber.
  - **Lever System:** A mechanism of levers connected to the float. As the float rises with the water level, the lever system magnifies this movement.
  - **Recording Pen:** A pen attached to the end of the lever system, which moves vertically.
  - **Clock-Driven Drum with Chart:** A cylindrical drum that rotates at a constant speed (e.g., one rotation per day or week) driven by a clockwork or electronic mechanism. A graph paper (chart) is wrapped around this drum. The x-axis of the chart represents time, and the y-axis represents the height of rainfall.
  - **Siphon Mechanism:** A siphon tube connected to the bottom of the float chamber. When the water level (and thus the float) reaches a predetermined maximum height (representing a specific amount of rainfall, e.g., 10 mm or 25 mm), the siphon automatically empties the float chamber quickly.
  - **Protective Housing:** An outer casing to protect the internal components from the weather.
- **Working Principle**
  - Rain falls into the collector and is directed into the float chamber.
  - As water accumulates, the water level in the float chamber rises, causing the float to rise.
  - The upward movement of the float is transmitted and magnified by the lever system, which moves the recording pen vertically upwards on the chart wrapped around the rotating drum.

- This traces a curve on the chart, with the height of the curve at any time indicating the accumulated rainfall. The slope of the curve indicates the intensity of rainfall.
- When the water level in the float chamber reaches the level of the siphon's inlet, the siphon action starts, rapidly emptying the chamber.
- As the chamber empties, the float falls, and the pen moves back to the zero line (or a lower point) on the chart.
- The process repeats with the next rainfall. The resulting trace on the chart provides a continuous record of rainfall over time.

• Installation

Similar to other rain gauges, the installation of a Float-Type Automatic Rainfall Recorder as per IMD specifications would involve:

- **Foundation:** A stable concrete or masonry foundation.
- **Leveling and Height:** The base of the gauge is fixed such that the rim of the collector is about 30 cm above the ground level and perfectly horizontal.
- **Exposure:** Open and level ground, away from obstructions (at least twice the height of the obstruction).
- **Protection:** Fencing to prevent damage.
- **Leveling of the instrument itself** is crucial for the proper functioning of the float and siphon mechanism.

Additionally, for an automatic recorder:

- The instrument needs to be placed such that the clock mechanism can be wound or the power supply is maintained.
- Easy access is required to change the chart paper regularly (e.g., daily or weekly).

• Recording Mechanism

The recording mechanism of a Float-Type Automatic Rainfall Recorder relies on the interplay between the rising water level and the rotating chart:

- **Float Movement to Pen Movement:** The height of the water in the float chamber directly corresponds to the vertical position of the recording pen on the chart. More rainfall means a higher float and thus a higher pen position.

- **Time Progression by Drum Rotation:** The clock mechanism drives the drum to rotate at a constant speed. This movement provides the time axis on the chart.
- **Continuous Record:** The combination of the pen's vertical movement (due to rainfall accumulation) and the drum's rotation creates a continuous graph of cumulative rainfall versus time. This allows for the determination of:
  - Total rainfall over a period.
  - The start and end time of rainfall events.
  - The intensity of rainfall (from the slope of the recorded curve).
- **Siphon Emptying:** The automatic siphoning provides a way to record rainfall beyond the maximum range of the float chamber. Each time the siphon empties, the pen drops, and the recording starts again from a lower level, creating a sawtooth-like pattern on the chart if rainfall is continuous. The total rainfall is then the sum of all the rises recorded.

While the Tipping Bucket Rain Gauge (TBRG) is more commonly used in modern automatic weather stations due to its digital output, the Float-Type gauge provides a direct graphical record, which can be valuable for visual analysis of rainfall patterns.

Precipitation & its Forms

1. **Which process is primarily responsible for precipitation?**

   a) Sublimation b) Condensation

   c) Evaporation d) Radiation

2. **Which of the following is not a form of precipitation?**

   a) Rain b) Dew

   c) Snow d) Hail

3. **What is the minimum size of a raindrop that reaches the ground?**

   a) 0.01 mm b) 0.1 mm

   c) 0.5 mm d) 2.0 mm

4. **Which form of precipitation is frozen water falling as ice crystals?**

   a) Sleet b) Snow

   c) Hail d) Freezing rain

5. **Which of the following is necessary for precipitation to occur?**

   a) Dry air b) Clear skies

   c) Condensation nuclei d) Evaporation

6. **Precipitation that evaporates before reaching the ground is known as:**
   a) Mist b) Fog
   c) Virga d) Drizzle
7. **Which instrument is used to measure rainfall?**
   a) Anemometer b) Thermometer
   c) Rain gauge d) Hygrometer
8. **Sleet is:**
   a) Frozen water vapor b) Snow mixed with rain
   c) Rain that freezes before hitting the ground d) Ice pellets formed directly in clouds
9. **Drizzle consists of water droplets usually smaller than:**
   a) 0.5 mm b) 1 mm
   c) 2 mm d) 3 mm
10. **Rain is categorized as 'heavy' when it exceeds:**
    a) 2 mm/hr b) 5 mm/hr
    c) 10 mm/hr d) 15 mm/hr
11. **What type of precipitation is formed by strong convective currents in cumulonimbus clouds?**
    a) Drizzle b) Hail
    c) Snow d) Sleet
12. **Freezing rain forms when:**
    a) Rain freezes before falling b) Rain falls through a freezing layer and freezes upon contact
    c) Snow melts completely d) Dew condenses and freezes
13. **What is the main difference between hail and sleet?**
    a) Composition b) Size
    c) Formation mechanism d) Color
14. **Which precipitation type typically occurs in summer thunderstorms?**
    a) Snow b) Sleet
    c) Hail d) Freezing rain
15. **Which precipitation results from warm air rising over cold air in a frontal system?**
    a) Convective rainfall b) Orographic rainfall
    c) Cyclonic rainfall d) Conduction rainfall
16. **Precipitation that occurs when moist air is lifted over a mountain is called:**
    a) Frontal b) Convective
    c) Orographic d) Cyclonic

17. Which form of precipitation forms during quiet, stable atmospheric conditions?

a) Showers
b) Hail
c) Drizzle
d) Thunderstorm rain

18. The Bergeron process is essential for:

a) Rain formation in the tropics
b) Cloud dispersion
c) Ice crystal growth in cold clouds
d) Formation of fog

19. Convective precipitation is associated with:

a) Mountains
b) Cold fronts
c) Rising warm air masses
d) Descending air

20. Rain that falls in short, intense bursts is called:

a) Showers
b) Drizzle
c) Fog
d) Mist

21. Cumulonimbus clouds are most often associated with:

a) Snow
b) Drizzle
c) Thunderstorms and heavy rain
d) Haze

22. Nimbostratus clouds typically bring:

a) Fair weather
b) Continuous precipitation
c) Dry air
d) Fog

23. Which cloud type indicates steady, prolonged rain?

a) Cirrus
b) Cumulus
c) Nimbostratus
d) Altocumulus

24. In cloud seeding, which chemical is commonly used to stimulate precipitation?

a) Potassium nitrate
b) Silver iodide
c) Sodium chloride
d) Ammonium sulfate

25. Which clouds are highest in the atmosphere?

a) Stratus
b) Cumulus
c) Cirrus
d) Nimbostratus

26. Warm clouds typically form precipitation by:

a) Bergeron process
b) Collision and coalescence
c) Sublimation
d) Radiation cooling

27. Which term refers to tiny droplets suspended in the air forming a cloud?

a) Vapor
b) Haze
c) Condensation droplets
d) Raindrops

28. Which cloud is associated with light drizzle or mist?

a) Stratocumulus
b) Stratus
c) Cirrus
d) Cumulonimbus

29. When precipitation forms due to temperature differences across a front, it is called:

a) Convective
b) Cyclonic or frontal
c) Orographic
d) Adiabatic

30. Cloud droplets must coalesce to reach a size sufficient for:

a) Dew formation
b) Air turbulence
c) Precipitation
d) Cloud thinning

31. Which device measures snow depth?

a) Snow gauge
b) Rain gauge
c) Barometer
d) Hygrometer

32. A pluviometer is used to measure:

a) Humidity
b) Wind speed
c) Rainfall
d) Snowfall

33. The unit of rainfall measurement is:

a) Celsius
b) Millimeters
c) Kilometers
d) Liters

34. Which precipitation form causes the most damage to crops?

a) Drizzle
b) Snow
c) Hail
d) Rain

35. Flooding is most commonly associated with:

a) Snow
b) Light rain
c) Heavy, prolonged rain
d) Mist

36. Which of the following can reduce visibility significantly?

a) Hail
b) Mist
c) Rain gauge
d) Haze

37. Rainwater harvesting is a strategy to:

a) Increase groundwater pollution
b) Reduce crop yield
c) Conserve water resources
d) Prevent hailstorms

38. Which region experiences monsoonal precipitation patterns?

a) Sahara Desert
b) Western Europe
c) Indian Subcontinent
d) Arctic

39. What type of precipitation do tropical cyclones produce?

a) Freezing rain
b) Hail
c) Torrential rain
d) Sleet

40. Which form of precipitation can occur even atabove-freezing temperatures near the ground?

a) Snow
b) Freezing rain
c) Sleet
d) Hail

41. Which condition is most favorable for snowfall?

a) Warm moist air
b) Sub-freezing temperatures and moist air
c) High pressure and heat
d) Dry air

42. What is 'graupel'?

a) Large hailstone
b) Snowflake coated with ice crystals
c) Rain droplet
d) Freezing mist

43. Which factor influences the type of precipitation received at ground level?

a) Cloud color
b) Wind direction
c) Vertical temperature profile
d) Solar radiation

44.During which season does convective precipitation occur most frequently in India?

a) Winter
b) Monsoon
c) Autumn
d) Spring

45.Which is a key factor in cloud droplet collision and growth?

a) Similar droplet size
b) Upward wind movement
c) High solar radiation
d) Clear sky

46. Which human activity can unintentionally influence precipitation patterns?

a) Cloud naming
b) Urbanization
c) Mountain climbing
d) Road building

47. Artificial precipitation is most often attempted during:

a) Floods
b) Cloudless days
c) Droughts
d) Snowstorms

48. Which atmospheric layer is most precipitation generated in?

a) Stratosphere
b) Troposphere
c) Mesosphere
d) Thermosphere

49. Droughts are defined primarily by the lack of:

a) Wind
b) Temperature
c) Precipitation
d) Solar radiation

50. Rainfall in India is predominantly caused by:

a) Westerlies
b) Monsoon winds
c) Trade winds
d) Local breezes

Answer Key (s)

1	b	2	b	3	c	4	b	5	c	6	c	7	c	8	c	9	a	10	b
11	b	12	b	13	c	14	c	15	c	16	c	17	c	18	c	19	c	20	a
21	c	22	b	23	c	24	b	25	c	26	b	27	c	28	b	29	b	30	c
31	a	32	c	33	b	34	c	35	c	36	b	37	c	38	c	39	b	40	b
41	b	42	b	43	c	44	b	45	b	46	b	47	c	48	b	49	c	50	b

7

Unraveling the Monsoon Rainfall in India and its Significance for Aquatic Ecosystems and Fisheries

Rainfall in India, is a phenomenon that is not just a meteorological event but the very lifeline of the subcontinent, profoundly shaping its landscapes, agriculture, and, most importantly for our study, its diverse aquatic ecosystems and the fisheries they support. India experiences a unique and dominant rainfall regime driven primarily by the **monsoon**, a seasonal reversal of wind patterns. The arrival, intensity, duration, and spatial distribution of this rainfall have far-reaching consequences for the availability of freshwater, the health of rivers, lakes, and wetlands, and the productivity of coastal and marine environments. For aspiring fisheries professionals in India, a thorough understanding of the patterns and vagaries of Indian rainfall is indispensable.

Consider the sheer diversity of aquatic habitats across India, from the snow-fed rivers of the Himalayas to the monsoonal wetlands of the Gangetic plains, the brackish backwaters of Kerala, and the vast coastline. The characteristics and productivity of each of these systems are intricately linked to the patterns of rainfall they receive. The monsoon replenishes rivers and groundwater, sustains agricultural practices that can impact water quality, influences the nutrient flow into coastal waters, and can even trigger crucial life cycle events for many aquatic species (Fig. 7.1).

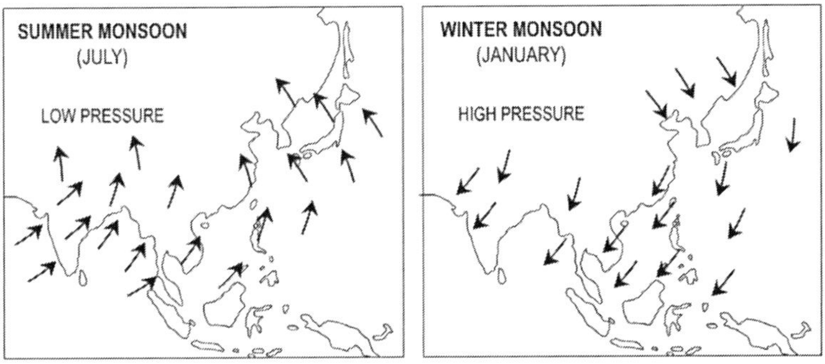

Fig. 7.1: Classical theory on monsoon

In this chapter, we will embark on a detailed exploration of rainfall in India, covering:

- **The mechanisms of the Indian Monsoon:** Understanding the atmospheric processes that lead to the seasonal onset and withdrawal of the monsoons (both the Southwest and Northeast monsoons).
- **Spatial and temporal distribution of rainfall:** Examining how rainfall varies across different regions of India and throughout the year, highlighting areas of high and low rainfall and the reasons behind these variations.
- **Types of rainfall in India:** Discussing orographic rainfall, convectional rainfall, and cyclonic rainfall, and their contribution to the overall precipitation.
- **The impact of rainfall on Indian aquatic ecosystems:** Analyzing how rainfall patterns influence river discharge, lake levels, groundwater recharge, wetland dynamics, and salinity in coastal areas.
- **The link between rainfall and Indian fisheries and aquaculture:** Exploring how the monsoon and its variability affect both inland and coastal fisheries, as well as aquaculture practices, including water availability, water quality, and the breeding cycles of important species.
- **Challenges related to rainfall variability:** Addressing issues such as droughts, floods, and their impact on aquatic resources and livelihoods dependent on them.

By the end of this chapter, you will gain a comprehensive understanding of the crucial role that rainfall plays in shaping India's aquatic environments and the implications for the sustainable management and development of its fisheries and aquaculture sectors. Let us delve into the intricacies of the Indian monsoon and its life-giving waters.

7.1 Onset of Southwest Monsoon

The onset of the Southwest Monsoon over India is a highly anticipated event, marking the transition from the hot, dry pre-monsoon season to a period of widespread rainfall. The India Meteorological Department (IMD) closely monitors various atmospheric and oceanic conditions to predict the date of its arrival.

7.1.1 Date of Onset

- **Normal Onset Date:** The normal date for the onset of the southwest monsoon over Kerala, which marks its entry into the Indian mainland, is

June 1st, with a standard deviation of about seven days (Fig.7.2).

- **First Point of Entry:** The monsoon typically first arrives over the **Andaman and Nicobar Islands** in the Bay of Bengal, usually around the third week of May (around May 15th to 20th). There's often a lag of about 10 days between its arrival in the Andaman Sea and the Kerala coast.
- **IMD's Prediction for 2025:** For the current year, **2025**, the IMD has predicted an **early onset** over Kerala, around **May 27th**, which is about five days earlier than the usual date.
- **Criteria for Onset Declaration:** The IMD declares the onset of the monsoon over Kerala based on a set of criteria that include:
  - **Rainfall:** At least 60% of 14 designated weather stations in Kerala and Lakshadweep reporting rainfall of 2.5 mm or more for two consecutive days after May 10th.
  - **Wind field:** Westerly winds up to a certain depth in the lower atmosphere and specific wind speeds.
  - **Outgoing Longwave Radiation (OLR):** A measure of energy emitted to space, which should be below a certain threshold, indicating increased cloudiness.

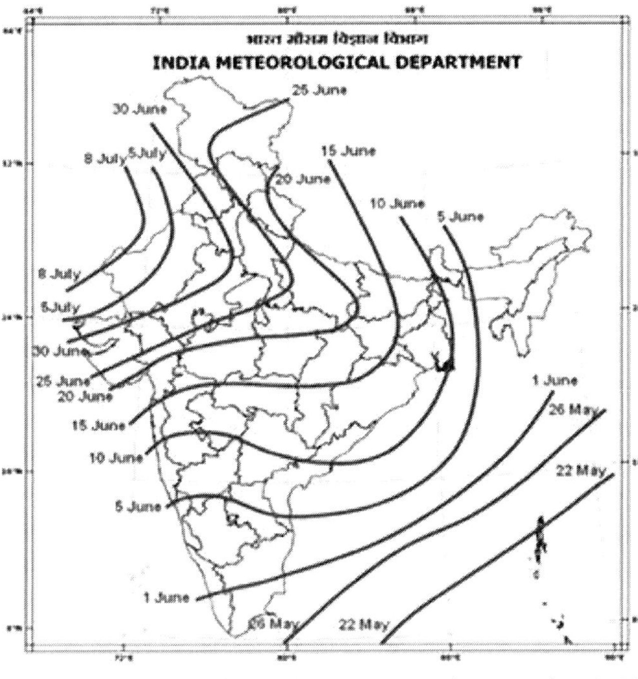

Fig. 7.2: New dates of normal monsoon onset/progress since 2020

7.1.2 How the Southwest Monsoon Spreads Across India

The southwest monsoon doesn't cover the entire country on a single day. It advances northwards and spreads across India in a phased manner, typically taking about 1 to 1.5 months to cover the entire landmass. The spread occurs in two main branches (Fig. 7.3):

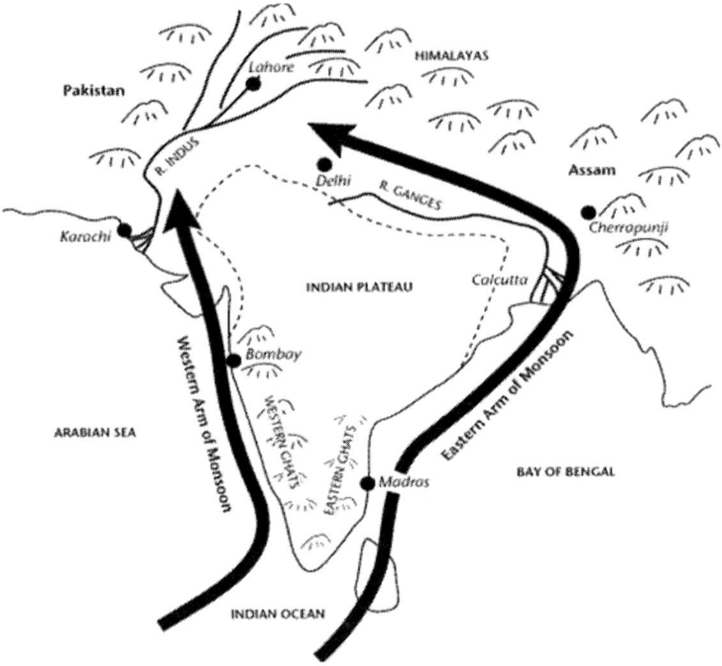

Fig. 7.3: Branches of SW monsoon

- **The Arabian Sea Branch**
  - This branch first strikes the **Western Ghats** along the southwestern coast of India (Kerala, Karnataka, Goa, Maharashtra). The steep slopes of the Western Ghats cause significant orographic rainfall on the windward side (western side).
  - It then progresses northwards along the west coast, reaching Mumbai generally by around June 10th.
  - Further inland, one stream of this branch moves towards the north-east, reaching Gujarat and eventually merging with the Bay of Bengal branch. Another stream moves along the foothills of the Himalayas.

- **The Bay of Bengal Branch**
  - This branch moves rapidly northeastwards across the Bay of Bengal, striking the coasts of **Myanmar and Bangladesh**, and then entering West Bengal and the northeastern states of India.
  - It brings heavy rainfall to Northeast India, including Meghalaya (home to the world's wettest places).
  - This branch then turns westwards along the Gangetic plains, gradually spreading across Bihar, Uttar Pradesh, and eventually merging with the Arabian Sea branch around Delhi by late June or early July.

Timeline of Spread (Normal Conditions)

- **End of May/Early June:** Onset over Kerala and parts of coastal Karnataka.
- **Mid-June:** Covers most of the west coast, parts of Maharashtra (including Mumbai), and advances into Northeast India. Kolkata usually receives it around June 7th.
- **Late June/Early July:** Spreads into central India, Gujarat, and reaches Delhi. The two branches (Arabian Sea and Bay of Bengal) merge.
- **Mid-July:** The monsoon typically covers the entire country by around July 15th.

The advance of the monsoon is often in "surges," with periods of rapid progress followed by slower phases or even temporary halts. The topography of the land, the presence of low-pressure systems, and the upper air circulation patterns all influence the speed and extent of its spread.

This year (2025), with the early onset predicted over Kerala, the subsequent spread across the country might also be slightly ahead of the normal schedule, provided there are no significant disruptions in the atmospheric conditions. The IMD will continue to issue updates on the progress of the monsoon as it advances across India.

7.2 Withdrawal of Southwest Monsoon

The withdrawal of the Southwest Monsoon over India is a more gradual and prolonged process compared to its relatively rapid onset. It begins as the land starts to cool down after the peak summer heat, leading to changes in pressure and wind patterns (Fig. 7.4).

7.2.1 Date of Withdrawal

- **Normal Start Date:** The withdrawal of the southwest monsoon typically begins from the extreme northwestern parts of India (e.g., Rajasthan) around the **first week of September**. The India Meteorological Department (IMD) usually sets **September 17th** as the normal date for the commencement of withdrawal from this region.
- **End Date:** The monsoon gradually recedes southwards, and it withdraws completely from the Indian landmass by around **mid-October**. The IMD considers **October 15th** as the normal final withdrawal date from the entire country.

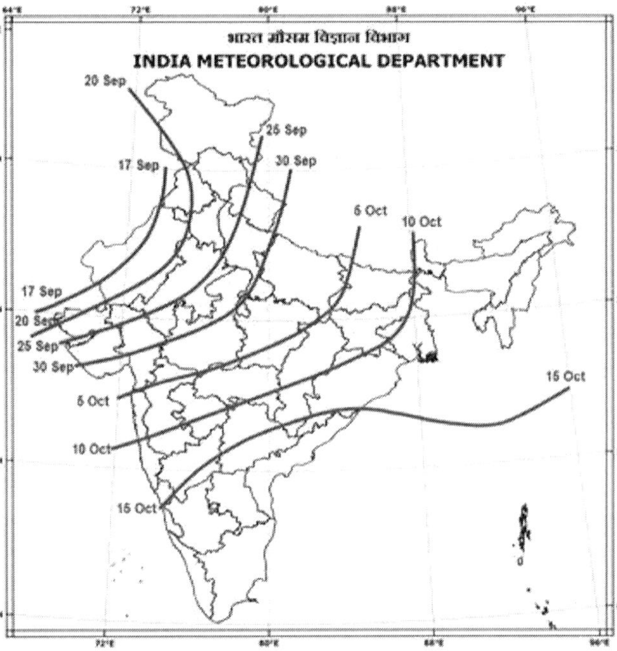

Fig. 7.4: New dates of normal withdrawal monsoon since 2020

Factors Influencing Withdrawal: The withdrawal is influenced by

- **Southward movement of the sun:** This leads to the southward shift of the low-pressure trough (associated with the monsoon) from the Gangetic plains.
- **Establishment of an anticyclone:** The development of an anticyclonic circulation in the lower troposphere over northwestern India, characterized by dry air and sinking motion, signals the end of the monsoon in that region.

- **Cessation of rainfall:** A continuous period of about five days with no significant rainfall in the northwestern parts.
- **Reduction in moisture:** A noticeable decrease in moisture content in the lower atmosphere, evident from satellite imagery.

7.2.2 How the Southwest Monsoon Recedes

The withdrawal of the southwest monsoon proceeds generally from northwest to southeast across India:

- **Northwest India (Early September):** The retreat begins from the northwestern states like Rajasthan, Punjab, Haryana, and parts of Gujarat. The air becomes drier, and rainfall ceases.
- **Northern Plains (Mid-September to Early October):** The withdrawal gradually extends to the remaining parts of North India, including Uttar Pradesh, Delhi, and Madhya Pradesh. The monsoon trough weakens and shifts southwards.
- **Central India (Late September to Mid-October):** The monsoon retreats from central parts of India.
- **Peninsular India (October):** The withdrawal from the peninsula is relatively faster. By mid-October, most parts of the peninsula experience a significant reduction in rainfall.
- **Extreme South (Mid to Late October):** The southwest monsoon finally withdraws from the southernmost parts of the peninsula. Around this time, the **Northeast Monsoon** begins to set in over the southeastern peninsula, bringing rainfall to Tamil Nadu, parts of Andhra Pradesh, and Kerala.

7.2.3 Timeline of Withdrawal (Normal Conditions)

- **Early September:** Withdrawal begins from West Rajasthan.
- **Mid-September:** Retreats from most of Northwest India.
- **End of September:** Withdraws from the northern plains and parts of Central India.
- **Mid-October:** Recedes from most of the Indian peninsula.
- **Late October:** Complete withdrawal from the entire country, with the onset of the Northeast Monsoon over the southeast peninsula.

The withdrawal phase is often characterized by clearer skies and a rise in daytime temperatures, while nights become cooler, particularly in North India. This period is sometimes referred to as "October heat" due to the combination of residual moisture and higher temperatures before the full establishment of the winter season.

7.3 Distribution of Southwest Monsoon Rainfall

The distribution of Southwest Monsoon rainfall across India is far from uniform, showcasing significant spatial variations due to a complex interplay of geographical factors, wind patterns, and the influence of topography.1Here's a detailed look:

7.3.1 Areas of High Rainfall (>200 cm annually during the SW Monsoon)

- **The Western Ghats (Windward Side):** The Arabian Sea branch of the monsoon strikes the Western Ghats perpendicularly. The orographic effect, where the moist air is forced to rise along the steep slopes, leads to heavy condensation and torrential rainfall on the western side. Some areas here receive over 250 cm of rainfall during this season.
- **Northeast India:** The Bay of Bengal branch of the monsoon flows into Northeast India, getting funneled by the hills. States like Meghalaya (including Mawsynram and Cherrapunji, some of the wettest places on Earth), Assam, Arunachal Pradesh, Nagaland, Manipur, Mizoram, and Tripura receive very high rainfall. The Khasi Hills in Meghalaya significantly enhance rainfall due to their funnel-like shape, trapping the moisture-laden winds.
- **Sub-Himalayan regions:** Areas along the foothills of the Himalayas also receive substantial rainfall as the Bay of Bengal branch moves westward and is forced to rise.

7.3.2 Areas of Medium Rainfall (100 - 200 cm annually during the SW Monsoon)

- **Southern parts of Gujarat:** Receives a moderate amount of rainfall from the Arabian Sea branch.
- **East Tamil Nadu:** While the southwest monsoon is not the primary rainy season for Tamil Nadu, it does receive some rainfall during this period.
- **Northeastern Peninsula:** This includes Odisha, Jharkhand, Bihar, and eastern Madhya Pradesh, which receive rainfall from the Bay of Bengal branch as it moves inland.
- **Northern Ganga Plain (along the sub-Himalayas):** As the Bay of Bengal branch moves westward.

- **Coastal Karnataka:** Receives good rainfall as part of the Arabian Sea branch's influence.

7.3.3 Areas of Low Rainfall (50 - 100 cm annually during the SW Monsoon)

- **Western Uttar Pradesh, Delhi, Haryana, Punjab:** These areas receive rainfall as the monsoon progresses westward, but the moisture content tends to decrease with distance from the Bay of Bengal. The merging of the Arabian Sea and Bay of Bengal branches also contributes.
- **Eastern Rajasthan:** Receives relatively less rainfall from the Arabian Sea branch as it moves parallel to the Aravalli Range. The Aravallis, running southwest-northeast, do not effectively block the monsoon winds to cause significant orographic lifting and rainfall in this region.
- **Gujarat (excluding the southern parts):** The northern parts and Kutch experience lower rainfall from the Arabian Sea branch, which moves parallel to the Aravallis.
- **Deccan Plateau:** Much of the interior Deccan Plateau receives less rainfall as it lies in the rain shadow region of the Western Ghats for the Arabian Sea branch. The monsoon winds lose much of their moisture by the time they cross the Ghats.

7.3.4 Areas of Inadequate Rainfall (< 50 cm annually during the SW Monsoon)

- **Parts of the Peninsula (especially leeward side of Western Ghats):** Regions in Maharashtra, Karnataka, and Andhra Pradesh that lie on the eastern side of the Western Ghats fall in the rain shadow zone and receive very scanty rainfall from the Arabian Sea branch. For example, Mahabaleshwar on the windward side receives over 600 cm, while Pune on the leeward side gets only around 70 cm.
- **Western Rajasthan:** This arid region receives very little rainfall from the Arabian Sea branch, which passes by without significant obstruction.
- **Ladakh:** Being a high-altitude cold desert, it receives minimal rainfall due to its location in the rain shadow of the Himalayas.

7.3.5 Factors Influencing the Distribution

- **Orographic Lift:** Mountain ranges like the Western Ghats and the Himalayas force moisture-laden winds to rise, causing heavy rainfall on the windward side. The leeward side experiences a rain shadow effect with much less rainfall.

- **Distance from the Sea:** Generally, rainfall decreases with increasing distance from the coast due to the gradual depletion of moisture in the winds.
- **Monsoon Depressions:** Low-pressure systems that form in the Bay of Bengal and the Arabian Sea move inland, significantly influencing the spatial distribution of rainfall along their track.
- **Wind Direction and Flow:** The direction of the monsoon winds and the patterns of air circulation determine which areas receive more direct and moisture-rich winds.
- **Topography:** Besides mountains, other topographical features can influence local rainfall patterns.
- **Location of the Monsoon Trough (ITCZ):** The position and movement of the monsoon trough across the Indo-Gangetic plains significantly affect rainfall distribution.

In summary, the distribution of southwest monsoon rainfall in India is highly varied, with the western coast and Northeast receiving the highest amounts, followed by the sub-Himalayan regions and parts of the central plains.14 Rainfall generally decreases as one moves westwards across the northern plains and into the leeward side of the Western Ghats and the arid regions of the northwest.

7.4 Break in the Monsoon

A **break in the monsoon** refers to a period during the Southwest Monsoon season (typically July and August) when there is a **significant reduction or complete cessation of rainfall over most parts of India**, particularly over the central and northern plains and the west coast. This dry spell occurs within the overall wet monsoon season, interspersed between periods of active rainfall. While the plains might experience a lull, rainfall often increases simultaneously along the Himalayan foothills, Northeast India, and sometimes parts of the southern peninsula.

Think of the monsoon not as continuous, uninterrupted rain, but as a period characterized by alternating wet spells and these drier "breaks."

The primary cause of a break in the monsoon is the **northward shift of the monsoon trough** (the low-pressure zone associated with the ITCZ) closer to the Himalayan foothills. Here's a breakdown of the mechanism:

7.4.1 *Normal Position of the Monsoon Trough*

Typically, during the peak monsoon months, the monsoon trough extends from

Rajasthan in the west to West Bengal in the east, across the Indo-Gangetic plains. This trough attracts moisture-laden winds from the Arabian Sea and the Bay of Bengal, leading to widespread rainfall across central and northern India.

- **Northward Shift of the Monsoon Trough:** When the monsoon trough shifts northwards and aligns itself along the foothills of the Himalayas, a significant change in the rainfall pattern occurs:
  - **Cessation of Rainfall over the Plains:** As the low-pressure system moves north, the plains of northern and central India experience a weakening of the monsoon circulation and a drastic reduction or complete stop in rainfall. The moisture-bearing winds are no longer effectively drawn over these areas.
  - **Increased Rainfall in the Himalayas and Northeast:** With the monsoon trough closer to the Himalayas, the monsoon winds are forced to ascend the steep slopes, leading to enhanced orographic rainfall in the Himalayan region and the northeastern states. This can sometimes result in floods in these areas.

7.4.2 Other factors that can contribute to breaks in the monsoon include

- **Absence of Monsoon Depressions:** The formation and movement of low-pressure systems or monsoon depressions over the Bay of Bengal are crucial for maintaining active monsoon conditions. If these depressions are infrequent or do not follow their usual track, it can lead to a break.
- **Westerly Winds over the West Coast:** When winds along the west coast become parallel to the coastline instead of perpendicular, they carry less moisture inland, leading to dry spells in the region.
- **Global Teleconnections:** Phenomena like El Niño-Southern Oscillation (ENSO) in the Pacific Ocean can influence the position and strength of the monsoon trough, indirectly contributing to break conditions.

7.4.3 Impact of Break Monsoons

Breaks in the monsoon can have significant and varied impacts:

- **Agriculture:** The most severely affected sector is agriculture, particularly rain-fed crops. Prolonged dry spells during crucial growth stages can lead to:
  - **Moisture stress in crops:** Wilting and stunted growth due to lack of water.

- **Reduced yields:** If the break is long and not followed by sufficient rainfall, it can significantly decrease agricultural productivity.
- **Delayed sowing for the next season:** If the withdrawal of the monsoon is also delayed after a long break, it can disrupt agricultural schedules.

- **Water Resources:** Breaks can lead to a halt in the replenishment of rivers, reservoirs, and groundwater, potentially leading to water scarcity if the break is prolonged.
- **Economy:** Given the dependence of the Indian economy on agriculture, a significant break in the monsoon that impacts crop yields can have broader economic repercussions, affecting rural incomes and overall growth.
- **Paradoxical Situations:** As mentioned earlier, while the plains suffer from a lack of rain, the Himalayan and northeastern regions might experience floods due to concentrated rainfall, leading to a paradoxical situation of drought in one part of the country and floods in another.
- **Ecosystems:** Prolonged dry spells can also impact natural vegetation and local ecosystems.

In summary, a break in the monsoon is a temporary but significant disruption in the typical rainfall pattern of the Southwest Monsoon, primarily caused by the northward shift of the monsoon trough. It leads to dry spells over central and northern India while often increasing rainfall in the Himalayan and northeastern regions, with potentially adverse impacts on agriculture, water resources, and the economy.

7.5 Northeast Monsoon

The **Northeast Monsoon**, also known as the **winter monsoon** or the **retreating monsoon**, is a distinct weather phenomenon affecting the Indian subcontinent, primarily during the months of **October to December**.While it contributes a smaller portion to India's overall annual rainfall compared to the Southwest Monsoon, it is critically important for the southeastern parts of the peninsula (Fig.7.5).

7.5.1 Formation of the Northeast Monsoon

The formation of the Northeast Monsoon is linked to the changes in atmospheric circulation that occur as the summer heating over the Indian landmass subsides and winter approaches:

- **Cooling of the Landmass:** After the Southwest Monsoon season, the Indian landmass begins to cool down more rapidly than the surrounding Indian Ocean. This cooling leads to the development of a **high-pressure area** over northern India.
- **Southward Shift of the ITCZ:** As the sun moves southwards towards the Tropic of Capricorn, the **Intertropical Convergence Zone (ITCZ)** also shifts south. The low-pressure trough associated with the monsoon weakens over the northern plains and gradually moves southward.
- **Reversal of Winds:** Due to the high pressure over the land and relatively lower pressure over the Bay of Bengal, the wind direction reverses. Cooler, drier air starts blowing outwards from the northeast across the land towards the sea. These are the **northeast trade winds**.
- **Moisture Pickup over the Bay of Bengal:** As these northeasterly winds travel over the warm waters of the Bay of Bengal, they pick up significant amounts of moisture. This moisture-laden air then brings rainfall to the eastern coast of South India.

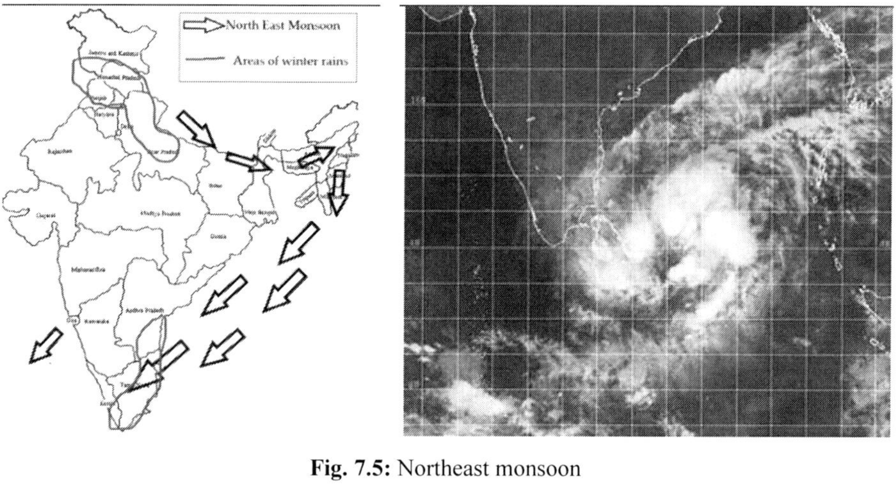

Fig. 7.5: Northeast monsoon

7.5.2 Spread of the Northeast Monsoon

Unlike the relatively well-defined onset and spread of the Southwest Monsoon from south to north, the onset of the Northeast Monsoon is more gradual and less dramatic. Its influence is largely confined to the southeastern peninsula:

- **Commencement:** The establishment of the northeasterly winds over the Bay of Bengal and the onset of rainfall along the Tamil Nadu coast usually marks the beginning of the Northeast Monsoon, typically around **mid-October**.

- **Limited Spread:** The influence of the Northeast Monsoon is primarily felt over **Tamil Nadu, coastal Andhra Pradesh, parts of Kerala, and south interior Karnataka**. It doesn't extend significantly into the interior of the country or the northern plains.
- **Rainfall Mechanism:** The rainfall associated with the Northeast Monsoon is often caused by:
  - **Easterly waves:** These are undulations in the tropical easterly flow that can bring rainfall.
  - **Low-pressure systems, depressions, and tropical cyclones:** The post-monsoon season (October-December) is the peak time for the formation of tropical cyclones in the Bay of Bengal. These systems can bring heavy and widespread rainfall to the eastern coast as they make landfall. Some of these cyclones can even cross the southern peninsula and emerge into the Arabian Sea.

7.5.3 Areas Under the Influence of the Northeast Monsoon

The Northeast Monsoon has a significant impact on the following regions:

- **Tamil Nadu:** This state is the **primary beneficiary** of the Northeast Monsoon, receiving about **40-60% of its annual rainfall** during this season. For many parts of Tamil Nadu, this is the main rainy season, crucial for agriculture (especially paddy, millets, and groundnut cultivation) and replenishing water reservoirs. The coastal districts often receive the heaviest rainfall, sometimes associated with cyclonic activity.
- **Coastal Andhra Pradesh:** This region also receives significant rainfall from the Northeast Monsoon, accounting for about **30-40% of its annual rainfall** in some areas, particularly the southern coastal districts. The rainfall is vital for agriculture and water management in this region.
- **Parts of Kerala:** While Kerala receives the bulk of its rainfall from the Southwest Monsoon, the southern parts of the state experience a **secondary rainy season** during the Northeast Monsoon. This rainfall is important for winter crops and maintaining water levels.
- **South Interior Karnataka:** The southeastern parts of Karnataka also receive some rainfall during the Northeast Monsoon. While not as high as in Tamil Nadu or coastal Andhra Pradesh, this rainfall is still beneficial for agriculture and water resources in this region.
- **Other Areas with Marginal Influence**

- **Rayalaseema (Andhra Pradesh):** Experiences some spillover rainfall from the Northeast Monsoon.
- **Puducherry and Karaikal:** Being geographically close to Tamil Nadu, these Union Territories also receive substantial rainfall during this season.

7.5.4 Areas Largely Unaffected

The Northeast Monsoon has minimal to no direct impact on most other parts of India, including the northern plains, central India, the west coast (except for some parts of Kerala), and Northeast India (which receives its main rainfall during the Southwest Monsoon).

In essence, the Northeast Monsoon is a crucial, albeit geographically limited, rainy season for the southeastern peninsula of India, particularly Tamil Nadu. It forms due to the reversal of winds after the summer monsoon, picking up moisture from the Bay of Bengal and bringing vital precipitation to these regions.

7.5.5 Distribution of Northeast Monsoon Rainfall

The **Northeast Monsoon** primarily influences the southeastern parts of the Indian peninsula during the months of **October, November, and December**. Let's break down its distribution with this timeframe in mind:

- **October**
  - As the Southwest Monsoon withdraws, the shift in wind patterns becomes noticeable. By mid-October, the northeasterly winds start establishing themselves over the Bay of Bengal.
  - **Tamil Nadu** and **coastal Andhra Pradesh** begin to experience an increase in rainfall. This is often the initial phase of their main rainy season. The rainfall during this month can be quite variable, sometimes starting with lighter showers and gradually intensifying.
  - **Kerala** and **south interior Karnataka** also start receiving some rainfall, marking the beginning of their secondary rainy season.
  - The formation of **tropical cyclones** in the Bay of Bengal becomes more frequent in October. These cyclones can bring heavy and widespread rainfall, along with strong winds, to the eastern coast, particularly affecting Tamil Nadu, Andhra Pradesh, and sometimes Odisha.

- **November**
  - November is typically the **peak month** for the Northeast Monsoon in terms of rainfall amount and its impact on the southeastern peninsula.
  - **Tamil Nadu** receives its highest rainfall during this month. Most parts of the state experience consistent and often heavy rainfall.
  - **Coastal Andhra Pradesh** also continues to receive significant rainfall, and the risk of cyclones impacting the coast remains high.
  - **Kerala** and **south interior Karnataka** continue to receive moderate rainfall.
  - The influence of the Northeast Monsoon generally does not extend far inland or to the northern parts of India during this period. The rest of the country experiences largely dry and pleasant weather as the transition to the winter season progresses.
- **December**
  - The intensity of the Northeast Monsoon starts to decrease in December.
  - Rainfall in **Tamil Nadu** and **coastal Andhra Pradesh** gradually reduces compared to November, but significant rainfall events can still occur, especially in the first half of the month.
  - **Kerala** and **south interior Karnataka** also see a further reduction in rainfall.
  - By the end of December, the Northeast Monsoon season effectively comes to an end, and the weather over the southeastern peninsula starts to become drier as the winter conditions become more established.

7.5.6 Summary of Regional Distribution over the Period

- **Tamil Nadu (October-December):** Receives the bulk (around 40-60%) of its annual rainfall, with November being the rainiest month. Coastal districts often get the heaviest.
- **Coastal Andhra Pradesh (October-December):** Experiences significant rainfall (30-40% of annual), with the risk of cyclonic rainfall being higher in October and November.
- **Kerala (October-December):** Receives its secondary rainy season, with moderate rainfall, mainly in the southern parts.
- **South Interior Karnataka (October-December):** Gets some rainfall, though less than the eastern coastal areas.
- **Rest of India (October-December):** Largely unaffected by the Northeast Monsoon, experiencing the transition towards the winter season with generally dry weather.

In essence, the distribution of the Northeast Monsoon's rainfall is heavily concentrated over the southeastern peninsula during October to December, with a peak in November. Tamil Nadu and coastal Andhra Pradesh are the primary beneficiaries, while Kerala and south interior Karnataka receive a lesser but still important amount. The influence diminishes significantly as one moves inland or northwards.

7.5.7 Cyclones and Depressions during Northeast Monsoon Season

The Bay of Bengal is a breeding ground for tropical disturbances, some of which intensify into depressions and cyclones, particularly during the **post-monsoon season (October-December)**, which aligns with the Northeast Monsoon. Several factors contribute to this: warm sea surface temperatures, lower vertical wind shear compared to the peak monsoon, and the remnants of typhoons from the Pacific sometimes entering the Bay of Bengal.

- **Depressions:** A **depression** is the first stage in the intensification of a tropical cyclone. It is characterized by:
  - **Low-pressure area:** A defined low-pressure center.
  - **Organized convection:** Thunderstorms that show some degree of organization around the center.
  - **Wind speed:** Maximum sustained surface winds up to **62 km/h (38 mph or 33 knots)**.

During the Northeast Monsoon season, depressions frequently form over the Bay of Bengal. While they are weaker than cyclones, they can still bring:

- **Heavy rainfall:** Capable of causing flooding in coastal and inland areas.
- **Gusty winds:** Leading to minor damage to weak structures and trees.
- **Rough seas:** Affecting fishing and maritime activities.

Many of the rainfall events associated with the Northeast Monsoon, especially in Tamil Nadu and Andhra Pradesh, are linked to the passage of these depressions. Some depressions intensify further into tropical cyclones.

- **Cyclones:** A **tropical cyclone** is a more intense weather system that develops from a tropical depression when certain favorable conditions persist. It is characterized by:
  - **Well-defined low-pressure center (the eye):** A region of relatively calm weather at the center.
  - **Organized and intense convection:** Bands of strong thunderstorms spiraling around the eye (eyewall being the most intense).

- **Wind speed:** Maximum sustained surface winds exceeding **63 km/h (39 mph or 34 knots)**. Depending on the intensity, they can be further classified into tropical storms, severe cyclonic storms, very severe cyclonic storms, and super cyclonic storms.

The post-monsoon season (October-December) is one of the **peak times for cyclone formation in the Bay of Bengal**. These cyclones can have a significant impact on the eastern coast of India and neighboring countries like Bangladesh and Myanmar.

7.5.8 Key aspects of Cyclones during the Northeast Monsoon

- **Frequency:** The Bay of Bengal experiences a higher frequency of tropical cyclones compared to the Arabian Sea. On average, a few cyclones form in the Bay of Bengal during this season.
- **Tracks:** Cyclones formed during this period often move west or northwest initially, posing a threat to the eastern coast of India (Tamil Nadu, Andhra Pradesh, Odisha, West Bengal). Some may recurve northeastwards.
- **Intensity:** These cyclones can range from tropical storms to very severe cyclonic storms, capable of causing widespread destruction.
- **Impact:** The impact can be severe, including:
    - **Storm surges:** Abnormal rise in sea level inundating coastal areas.
    - **High winds:** Causing damage to infrastructure, homes, and vegetation.
    - **Torrential rainfall:** Leading to widespread flooding.

7.5.9 Examples (Hypothetical for the 2025 Northeast Monsoon)

Given the current date (May 15, 2025), we cannot provide specific examples of cyclones that will occur during the upcoming Northeast Monsoon. However, historically, this period has seen the formation of significant cyclones. For instance, in November 2024, Cyclone Michaung impacted Tamil Nadu and Andhra Pradesh.

In the upcoming Northeast Monsoon (October-December 2025), it is plausible that:

- Several low-pressure systems and depressions will form over the Bay of Bengal, bringing periods of enhanced rainfall to the southeastern peninsula.
- At least one or two of these systems might intensify into tropical cyclones, potentially making landfall along the eastern coast, bringing

the associated hazards of strong winds, heavy rain, and storm surges to vulnerable coastal districts.

The India Meteorological Department (IMD) closely monitors the Bay of Bengal for the development of such systems and issues timely warnings.

In summary, the Northeast Monsoon season in India is characterized by the formation of depressions and cyclones in the Bay of Bengal, which are crucial for the rainfall in the southeastern peninsula but also pose a significant risk of severe weather events along the eastern coastline.

7.6 Distribution of Annual Rainfall in India

The average annual rainfall of India is about **118 cm**, but this figure hides a vast disparity in its distribution across different regions. Several factors such as topography, distance from the sea, prevailing wind direction, and the location of mountain ranges significantly influence how much rainfall different parts of India receive. Here's a detailed look at the distribution of annual rainfall in India:

7.6.1 Zones of Very High Rainfall (Above 200 cm annually)

- **The Western Ghats (Windward Side):** The western slopes of the Western Ghats, intercepting the moisture-laden Arabian Sea branch of the Southwest Monsoon, receive exceptionally heavy rainfall, often exceeding **250 cm**.Areas like the Konkan coast, parts of Karnataka and Kerala within the Ghats experience this. The orographic effect is the primary reason here.
- **Northeast India:** This region, comprising states like Meghalaya, Assam, Arunachal Pradesh, Nagaland, Manipur, Mizoram, and Tripura, receives very high rainfall primarily from the Bay of Bengal branch of the Southwest Monsoon.
  - **Meghalaya:** Contains some of the wettest places on Earth, such as **Mawsynram** and **Cherrapunji**, which can receive over **1000 cm** of rainfall annually. The unique funnel shape of the Khasi Hills enhances the concentration of moisture-laden winds.
  - **Sub-Himalayan West Bengal and Sikkim:** These areas also receive high rainfall due to their proximity to the Himalayas, which force the monsoon winds to rise.

7.6.2 Zones of High Rainfall (100 - 200 cm annually)

- **Eastern India:** This includes parts of West Bengal, Odisha, Bihar, and Jharkhand, which receive substantial rainfall from the Bay of Bengal branch of the Southwest Monsoon as it moves inland.

- **Coastal Plains:** Areas along the western and eastern coastal plains (excluding Tamil Nadu coast during the SW monsoon) generally receive rainfall in this range.
- **Parts of the Northern Plains:** Areas along the sub-Himalayan belt of the northern Ganga plain also fall into this category.
- **Southern Gujarat:** Receives a good amount of rainfall from the Arabian Sea branch.

7.6.3 Zones of Moderate Rainfall (50 - 100 cm annually)

- **Central India:** Much of Madhya Pradesh and parts of Chhattisgarh receive moderate rainfall.
- **Deccan Plateau:** The rain shadow region to the east of the Western Ghats, covering parts of Maharashtra, Karnataka, and Andhra Pradesh, receives comparatively less rainfall.
- **Parts of the Northern Plains:** Western Uttar Pradesh, Haryana, and Punjab receive rainfall in this range as the monsoon weakens moving westward.
- **Eastern Rajasthan:** Receives less rainfall as the Arabian Sea branch moves parallel to the Aravalli Range, offering little obstruction.

7.6.4 Zones of Low Rainfall (Less than 50 cm annually)

- **Western Rajasthan:** The arid and semi-arid regions of western Rajasthan receive very scanty rainfall, often less than **25 cm**. The Arabian Sea branch of the monsoon flows parallel to the Aravalli hills, and there are no significant barriers to cause orographic lift.
- **Parts of Gujarat:** The northern parts of Gujarat, including the Kutch region, also receive low rainfall.
- **Leeward side of the Western Ghats:** A significant area of the Deccan Plateau falls in the rain shadow region of the Western Ghats and thus receives very little rainfall.
- **Ladakh:** Being a cold desert situated in the rain shadow of the Himalayas, Ladakh receives extremely low rainfall.

7.6.5 Factors Explaining the Uneven Distribution

- **Topography:** The presence of mountain ranges like the Western Ghats and the Himalayas plays a crucial role in intercepting monsoon winds and causing orographic rainfall on the windward side, while creating rain shadow regions on the leeward side.

- **Distance from the Sea:** Coastal areas generally receive more rainfall than the interior regions due to their proximity to moisture sources.
- **Direction of Monsoon Winds:** Areas that lie directly in the path of the moisture-laden monsoon winds receive more rainfall. For example, the southwest monsoon winds directly hit the Western Ghats and Northeast India.
- **Influence of Ocean Currents:** Sea surface temperatures influenced by ocean currents can affect the amount of moisture in the air.
- **Low-Pressure Systems:** The formation and movement of monsoon depressions and cyclones in the Bay of Bengal and the Arabian Sea significantly influence the distribution of rainfall. Areas along their track receive heavier rainfall.

In conclusion, the distribution of annual rainfall in India is highly uneven, ranging from some of the wettest places on Earth to arid deserts, all within the same country. This diversity is primarily due to the complex interplay of geographical and meteorological factors that influence the behavior of the monsoon winds and the occurrence of precipitation.

7.7 Impact of Rainfall on Indian Aquatic Systems

The impact of rainfall on Indian aquatic systems is multifaceted, with both positive and negative consequences. Here's a detailed look:

7.7.1 Positive Impacts

- **Replenishment of Water Bodies:** The monsoon rains are crucial for replenishing rivers, lakes, ponds, and groundwater, which are essential for aquatic life.
- **Habitat Creation:** Rainfall can create temporary water bodies and expand existing ones, providing breeding grounds and habitats for various aquatic organisms like amphibians and some fish species.
- **Nutrient Input:** Runoff from rainfall can carry essential nutrients from land into aquatic systems, potentially boosting primary productivity.
- **Spawning Cues:** The onset of the monsoon and associated rainfall and water level changes can act as environmental cues for the spawning of many fish species, including Indian Major Carps.
- **Increased Fish Availability (in some contexts):** Increased water levels can lead to the inundation of new areas, providing more space and potentially more food for fish, at least temporarily.

7.7.2 Negative Impacts

- **Flooding and Habitat Destruction:** Excessive rainfall can lead to devastating floods, destroying aquatic habitats, displacing or killing aquatic life, and damaging aquaculture infrastructure.
- **Water Quality Changes:** Heavy rainfall and runoff can carry pollutants (sediments, pesticides, fertilizers, sewage) into water bodies, degrading water quality and harming aquatic organisms. Increased turbidity can reduce sunlight penetration, affecting primary production.
- **Changes in Salinity:** In coastal areas and estuaries, heavy rainfall can drastically reduce salinity, which can stress or kill marine and brackish water species.
- **Disease Outbreaks:** Standing water from heavy rains can become breeding grounds for disease vectors that can affect both wild and farmed aquatic populations. Wet conditions can also promote fungal and other infections.
- **Disruption of Breeding Cycles:** While the monsoon triggers spawning for some, unpredictable or extreme rainfall can wash away nests or disrupt breeding behaviors.
- **Impact on Aquaculture:** Heavy rains can cause the loss of stock from ponds due to flooding, introduce unwanted species, alter water parameters like temperature and pH, and increase the risk of disease. Reduced sunlight during prolonged rainfall can also affect algal growth, which is important in some aquaculture systems.

In summary, rainfall, particularly the monsoon, is a dominant force shaping Indian aquatic systems. While it is essential for maintaining water availability and triggering life cycles, its variability and intensity can also pose significant challenges, leading to habitat destruction, water quality degradation, and disruptions in both wild fisheries and aquaculture.

7.8 Thunderstorms

A thunderstorm is a localized storm associated with lightning and thunder, typically accompanied by heavy rainfall, strong winds, and sometimes hail. They are formed by the rapid upward movement of warm, moist air into colder levels of the atmosphere. This process, known as **convection**, leads to the formation of cumulonimbus clouds, which are the birthplace of thunderstorms.

Three basic ingredients are needed for thunderstorm formation:

- **Moisture:** To form clouds and precipitation.
- **Unstable Air:** Warm air near the surface that is buoyant and tends to rise.
- **A Lifting Mechanism:** Something to initiate the upward motion (e.g., heating of the ground, mountains, weather fronts).

7.8.1 Types of Thunderstorms

Thunderstorms in India vary based on their origin and characteristics. Some common types include:

- **Convectional Thunderstorms (Thermal Thunderstorms):** These develop due to intense heating of the ground during the summer months (especially April-May). The warm air rises, cools, and condenses, leading to the formation of cumulonimbus clouds. These are often localized and short-lived. In some parts of India, they are known by local names like "Mango Showers" (Kerala) and "Blossom Showers" (Karnataka).
- **Orographic Thunderstorms:** These occur when warm, moist air is forced to rise as it encounters a mountain barrier. As the air rises and cools on the windward side, condensation and precipitation occur. These are common in regions like the Himalayas (including Jammu & Kashmir) and the Western Ghats. "Cloud bursts," which can lead to flash floods, are often associated with orographic lifting.
- **Frontal Thunderstorms:** These are associated with weather fronts, particularly cold fronts, where a mass of cold air pushes under a mass of warm, moist air, forcing it to rise. These can be more organized and widespread than convectional thunderstorms.
- **Nor'westers (Kalbaisakhi):** These are severe thunderstorms that occur during the pre-monsoon season (March-May) over eastern and northeastern India, particularly West Bengal, Jharkhand, Bihar, Orissa, and the northeastern states. They typically move from the northwest to the southeast and are known for their strong winds, heavy rainfall, and sometimes hail.
- Based on lifespan and organization, thunderstorms can also be classified as single-cell, multi-cell, squall lines (a line of multiple thunderstorm cells), and supercells (large, long-lived storms with a rotating updraft, capable of producing severe weather like intense tornadoes, though supercells are less common in India compared to some other parts of the world).

7.8.2 Impact of Thunderstorms on Indian Aquatic Systems and Fisheries

Thunderstorms can have both direct and indirect impacts

- **Heavy Rainfall and Flooding:** Intense rainfall from thunderstorms can lead to flash floods, especially in urban areas and hilly regions. This can damage aquaculture infrastructure (ponds, cages), wash away fish stocks, and alter the physical habitat of wild aquatic species. Increased river discharge can also affect estuarine environments by changing salinity.
- **Water Quality:** Runoff associated with thunderstorms can carry pollutants, sediments, and nutrients into water bodies, affecting water quality parameters like turbidity, dissolved oxygen, and nutrient levels, which can stress or harm aquatic life.
- **Wind and Waves:** Strong winds generated by thunderstorms can create large waves on lakes and coastal areas, potentially damaging aquaculture structures and disrupting fishing activities.
- **Lightning:** While direct strikes on water bodies might have localized impacts, lightning can pose a risk to people working in and around water, including fishermen and aquaculture farmers. It can also cause power outages affecting aquaculture facilities that rely on electricity for pumps and other equipment.
- **Hail:** Hail associated with severe thunderstorms can directly injure or kill fish in aquaculture ponds, especially smaller or surface-dwelling species.

7.8.3 Regional Variations in Thunderstorm Activity

Thunderstorm activity varies across India due to differences in topography, climate, and synoptic conditions.

- **Northeast India:** This region experiences a high frequency of thunderstorms, particularly during the pre-monsoon and monsoon seasons, due to the convergence of moist winds and the influence of the Himalayas.
- **Eastern India:** States like West Bengal and Odisha also experience significant thunderstorm activity, including the severe Nor'westers.
- **Northwest India:** While generally drier, this region can experience dust storms and thunderstorms, especially during the summer.
- **Peninsular India:** Thunderstorms are common, particularly during

the monsoon season, influenced by localized heating and moisture availability. Kerala tends to have a high frequency of thunderstorms during the monsoon.

7.8.4 Hazards Associated with Thunderstorms

Thunderstorms can be hazardous due to:

- **Lightning:** A major cause of weather-related fatalities in India.
- **Heavy Rainfall and Flash Flooding:** Leading to loss of life, damage to property, and disruption of infrastructure.
- **Strong Winds:** Capable of causing structural damage and uprooting trees.
- **Hail:** Can damage crops, property, and harm livestock and aquatic life.
- **Downbursts:** Localized columns of sinking air within a thunderstorm that can produce damaging straight-line winds.

Understanding thunderstorms in India, their formation, types, and impacts, is crucial for those involved in fisheries and aquaculture to prepare for and mitigate potential risks to their livelihoods and the aquatic environments they depend on.

7.9 Cloudiness and Ceiling

Let's discuss **cloudiness** and **ceiling** in detail, focusing on their meteorological significance and potential relevance to fisheries and aquaculture.

7.9.1 Cloudiness

It refers to the fraction of the sky that is obscured by clouds when observed from a particular location. It's a measure of how much of the sky is covered by clouds and is usually reported in terms of octas (eighths) or as a percentage.

- **Clear Sky:** 0 octas (0% cloud cover)
- **Few:** 1-2 octas (12.5-25% cloud cover)
- **Scattered:** 3-4 octas (37.5-50% cloud cover)
- **Broken:** 5-7 octas (62.5-87.5% cloud cover)
- **Overcast:** 8 octas (100% cloud cover)

- **Significance of Cloudiness**
  - **Solar Radiation:** Cloud cover significantly affects the amount of solar radiation (insolation) reaching the Earth's surface and water bodies. More clouds mean less direct sunlight. This has implications for:

    - **Water Temperature:** Reduced solar radiation leads to cooler surface water temperatures.
    - **Primary Productivity:** Less sunlight can limit photosynthesis by phytoplankton and aquatic plants.
    - **Aquaculture:** Sunlight intensity can affect the growth of algae in ponds and the overall thermal conditions.
- **Air Temperature:** Clouds also influence air temperature. During the day, they reflect some incoming solar radiation, leading to cooler temperatures. At night, they can trap outgoing longwave radiation, resulting in warmer temperatures compared to a clear night. This can indirectly affect water temperatures.
- **Precipitation:** Clouds are a prerequisite for most forms of precipitation (rain, snow, hail). The type and amount of cloud cover can give clues about the likelihood of precipitation.
- **Visibility:** Dense cloud cover, especially low clouds and fog (which is essentially a cloud at ground level), can significantly reduce visibility, impacting navigation for fishing vessels.

7.9.2 Ceiling

It is defined as the height above the Earth's surface of the lowest layer of clouds reported as "broken" (more than half the sky covered) or "overcast" (the entire sky covered). It essentially refers to the lowest cloud layer that obscures more than half of the sky.

- If the sky is clear, few, or scattered, there is no ceiling. In such cases, the report might indicate "clear" or specify the amount of cloud cover without a ceiling height.
- The ceiling is usually reported in feet or meters above ground level (AGL).

- **Significance of Ceiling**
    - **Aviation:** Ceiling is critically important for aviation safety, as it determines whether visual flight rules (VFR) can be followed. Low ceilings can necessitate instrument flight rules (IFR).
    - **General Weather Conditions:** A low ceiling often indicates the presence of significant cloud cover and can be associated with poor weather conditions, potentially including precipitation and reduced visibility.

- **Indirectly for Fisheries:** While ceiling doesn't directly impact aquatic systems as much as cloudiness or precipitation, a persistently low ceiling might be associated with prolonged periods of overcast skies, reducing solar radiation input to water bodies. It can also be an indicator of broader weather systems that could affect fishing operations. For example, a low ceiling associated with a developing storm would imply potential hazards at sea or for aquaculture.

In summary

- **Cloudiness** describes how much of the sky is covered by clouds and significantly influences solar radiation reaching aquatic environments.
- **Ceiling** describes the height of the *lowest* substantial cloud layer (broken or overcast) and is more critical for activities like aviation but can also indicate the presence of weather systems relevant to those working on or near water.

For fisheries students in India, understanding cloudiness helps in assessing light availability for primary production in aquaculture ponds and natural water bodies, as well as anticipating changes in water temperature due to variations in solar heating. Ceiling might be more relevant in the context of navigating fishing vessels, especially if poor weather with low clouds and reduced visibility is anticipated.

Unraveling the Monsoon: Rainfall in India and its Significance for Aquatic Ecosystems and Fisheries

1. **Which oceanic phenomenon strongly influences the Indian monsoon?**
   a) El Niño b) La Niña
   c) Indian Ocean Dipole d) All of the above
2. **The Indian monsoon originates primarily due to:**
   a) Jet streams b) Differential heating of land and ocean
   c) Ocean currents d) Polar vortex
3. **Which jet stream is associated with the onset of the Indian monsoon?**
   a) Polar jet stream b) Western jet stream
   c) Tropical easterly jet d) Subtropical jet
4. **What is the usual onset period of the southwest monsoon in Kerala?**
   a) May 15 b) June 1
   c) July 1 d) June 15

5. The retreating monsoon primarily brings rainfall to:

a) Northern India
b) North-eastern India
c) Central India
d) Tamil Nadu

6. Break monsoon refers to:

a) Total cessation of rainfall
b) Periods of low rainfall during active monsoon
c) Withdrawal of monsoon
d) Start of winter rains

7. Which region of India receives the highest rainfall due to the monsoon?

a) Western Rajasthan
b) Kerala
c) Mawsynram
d) Gujarat

8. Which of these affects the spatial variability of monsoon rainfall?

a) Orography
b) Latitude
c) Vegetation cover
d) All of the above

9. Which instrument is used to measure rainfall?

a) Hygrometer
b) Rain gauge
c) Barometer
d) Pyranometer

10. The monsoon trough is a:

a) High-pressure zone
b) Belt of dry winds
c) Low-pressure zone across northern India
d) Area of fog formation

11. Monsoon rainfall is essential for:

a) Recharging groundwater
b) Filling reservoirs and lakes
c) Maintaining stream flows
d) All of the above

12. Increased turbidity during monsoon affects fish by:

a) Enhancing breeding
b) Improving oxygen concentration
c) Reducing light penetration and photosynthesis
d) Increasing salinity

13. Flooding during intense monsoon rains can lead to:

a) Fish migration
b) Fish mortality due to oxygen depletion
c) Spread of invasive species
d) All of the above

14. Sudden inflow of freshwater in estuaries during monsoon:

a) Increases salinity
b) Disrupts fish osmoregulation
c) Enhances coral growth
d) Causes water stratification

15. Rainfall increases primary productivity in aquatic systems by:

a) Lowering pH
b) Contributing nutrients through runoff
c) Increasing salt levels
d) None of the above

16. Which fish species thrives during monsoon breeding?

a) Catla
b) Rohu
c) Hilsa
d) All of the above

17. Rain-fed ponds are mostly used for:

a) Mariculture
b) Brackish water aquaculture
c) Freshwater aquaculture
d) Trout farming

18. Excessive rainfall in hatcheries may lead to:

a) Optimum fertilization
b) Dilution of water quality
c) Higher hatching rate
d) Reduction in parasitic load

19. Monsoon failure can lead to:

a) Prolonged fish breeding
b) Enhanced water productivity
c) Shrinking aquatic habitats
d) None of the above

20. Which state's inland fisheries are most dependent on monsoon rainfall?

a) Rajasthan
b) Punjab
c) West Bengal
d) Himachal Pradesh

21. IMD issues monsoon forecast using:

a) Empirical model
b) Statistical model
c) Dynamical model
d) All of the above

22. A good monsoon forecast helps fisheries by:

a) Deciding breeding schedules
b) Managing stocking density
c) Planning pond preparation
d) All of the above

23. Early warning systems for floods are crucial for:

a) Coastal aquaculture
b) Inland fisheries
c) Fish seed farms
d) All of the above

24. Which satellite aids in monsoon observation?

a) INSAT-3D
b) CartoSat
c) RISAT
d) Oceansat

25. Climate-resilient aquaculture practices include:

a) Polyculture
b) Water harvesting
c) Raised pond dykes
d) All of the above

26. Which scheme supports monsoon-dependent aquaculture farmers in India?

a) PMFBY
b) PMMSY
c) RKVY
d) MGNREGA

27. A deficit monsoon may prompt government action in:

a) Subsidies for fish feed
b) Fish insurance schemes
c) Emergency seed stocking
d) All of the above

28. Watershed management improves aquaculture by:

a) Increasing water storage
b) Preventing runoff pollution
c) Promoting integrated farming
d) All of the above

29. The role of gram panchayats during monsoon in aquaculture includes:

a) Stocking fish
b) Monitoring flood-prone ponds
c) Issuing cyclone alerts
d) All of the above

30. Monsoon-related fisheries disaster is best mitigated by:

a) Overfeeding
b) Crop rotation
c) Early warning and response systems
d) Genetic modification

31. Which part of India receives significant rainfall from both southwest and northeast monsoons?

a) Punjab
b) Rajasthan
c) Tamil Nadu
d) Madhya Pradesh

32. The Western Ghats receive high rainfall due to:

a) Orographic lifting of moist winds
b) Convective rainfall
c) Thunderstorm activity
d) Local humidity accumulation

33. Bay of Bengal branch of monsoon first strikes:

a) Gujarat
b) Kerala
c) West Bengal
d) Odisha

34. Which region experiences the monsoon withdrawal first?

a) Rajasthan
b) Tamil Nadu
c) Gujarat
d) Kerala

35. Seasonal inland water bodies in India are mostly filled by:

a) Snowmelt
b) Irrigation canals
c) Monsoon rains
d) Groundwater pumping

36. Monsoon-driven changes in water pH are caused by:

a) Rainwater acidity
b) Runoff of organic matter
c) Photosynthetic activity
d) All of the above

Rainfall increases dissolved nutrients in aquatic systems by:

a) Increasing solar radiation
b) Reducing wind velocity
c) Introducing allochthonous materials
d) Raising atmospheric pressure

37. Fish kills after heavy rainfall are often caused by:

a) Overfeeding
b) pH stabilization
c) Oxygen depletion
d) Sediment settling

38. Increased flow during monsoon helps in:

a) Eutrophication
b) Siltation
c) Spawning migration of riverine species
d) Temperature stratification

39. Which aquatic plant growth is stimulated by monsoon nutrient inflow?

a) Azolla
b) Lemna
c) Phytoplankton
d) All of the above

41. A sudden drop in water temperature due to monsoon showers can:

a) Improve breeding
b) Increase oxygen solubility
c) Stress fish metabolism
d) Make water saline

42. Prolonged cloud cover during monsoon affects:

a) Water clarity
b) Photosynthetic rate
c) Dissolved oxygen levels
d) All of the above

43. Fish seed transportation is more difficult during monsoon due to:

a) Poor water quality
b) Road inaccessibility and flooding
c) Fish maturity
d) Lack of packaging

44. During flash floods, fish ponds must be protected by:

a) Raising dykes
b) Adding chlorine
c) Draining water completely
d) Harvesting early

45. Seasonal fish species that benefit from monsoon flooding include:

a) Snakeheads
b) Eels
c) Climbing perch
d) All of the above

46. Sustainable aquaculture planning should consider:

a) Monsoon onset and intensity
b) Stocking density and timing
c) Water retention structures
d) All of the above

47. Monsoon variability due to climate change may lead to:

a) Early spawning
b) Frequent pond drying
c) Predictable rainfall
d) Increase in invasive species

48. Artificial ponds supported by rainwater harvesting during monsoon are called:

a) Tidal tanks
b) Percolation tanks
c) Catchment basins
d) Rain-fed reservoirs

49. Monitoring monsoon for fisheries planning should include:

a) Rainfall forecasts
b) Soil moisture
c) Reservoir inflow data
d) All of the above

50. Best practice during delayed monsoon includes:

a) Early stocking of fingerlings
b) Complete water change
c) Reducing feed input and monitoring water
d) Increasing pond depth

Answer Key (s)

1	d	2	b	3	c	4	b	5	d	6	b	7	b	8	d	9	b	10	c
11	d	12	c	13	d	14	b	15	b	16	d	17	c	18	b	19	c	20	c
21	d	22	d	23	d	24	a	25	d	26	b	27	d	28	d	29	b	30	c
31	c	32	a	33	c	34	a	35	c	36	d	37	c	38	c	39	c	40	d
41	c	42	d	43	b	44	a	45	d	46	d	47	b	48	d	49	d	50	c

8

Atmospheric Pressure and Winds

Imagine the air around us. Even though we can't see it, air has weight. Atmospheric pressure is essentially the **force exerted by the weight of the air molecules above a particular point on the Earth's surface.**

Think of it like this: Imagine you're at the bottom of a swimming pool. The water above you presses down, and you feel that pressure. Similarly, the atmosphere, a giant ocean of air, has layers of air molecules, and their weight presses down on everything below, including us.

- **Definition:** Atmospheric pressure is defined as the **force per unit area exerted by the weight of the air column extending from the surface of the earth to the top of the atmosphere above that area** (Fig.8.1).

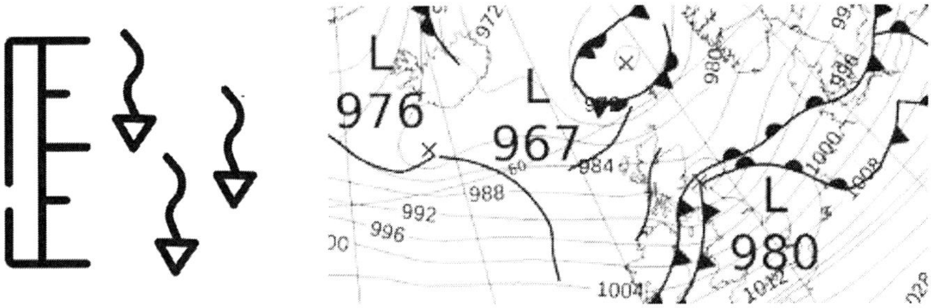

Fig. 8.1: Atmospheric pressure

- **How to Explain It Simply**
  - **Air has weight:** Just like water or a pile of books, the air surrounding the Earth has mass, and thus, weight due to gravity.
  - **Weight creates pressure:** This weight of the air presses down on the surface. The more air there is above a certain point, the greater the pressure.
- **Like a stack:** Imagine a stack of blankets. The blanket at the bottom has to support the weight of all the blankets above it, so it experiences more pressure. Similarly, the air at lower altitudes has the weight of all the air above pressing down on it.

The most common formula used to calculate atmospheric pressure, particularly when considering a column of fluid is:

$P=\rho gh$

Where:

- **P** = Atmospheric pressure. This is the force per unit area exerted by the weight of the air. It's typically measured in Pascals (Pa), but other units like atmospheres (atm), millibars (mb), or inches of mercury (inHg) are also used.
- ρ (rho) = Density of the fluid. In the case of atmospheric pressure measured by a barometer, the fluid is often mercury, so you'd use the density of mercury. For calculations involving air itself, you'd use the density of air. Density is measured in kilograms per cubic meter (kg/m^3).
- **g** = Acceleration due to gravity. This is the constant acceleration of an object falling freely near the Earth's surface. Its standard value is approximately 9.81 meters per second squared (m/s^2).
- **h** = Height of the fluid column. This is the vertical height of the column of fluid (mercury, in a barometer) that is being supported by the atmospheric pressure. It's measured in meters (m).

In simpler terms, this formula tells us that the pressure at the bottom of a column of fluid is directly proportional to the density of the fluid, the acceleration due to gravity, and the height of the column

8.1 Measurement of Atmospheric Pressure

The instrument used to measure atmospheric pressure is called Barometer. In meteorology, the atmospheric pressure is measured using an instrument called Fortin's Barometer (Fig.8.2).

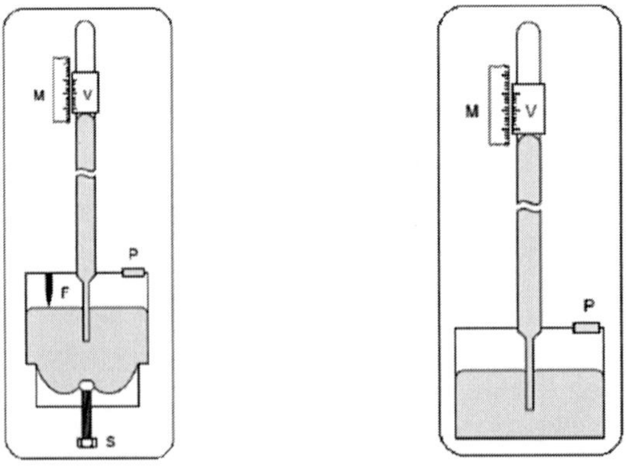

Fig. 8.2: Fortin barometer to measure atmospheric pressure

The Fortin barometer is a type of mercury barometer used for precise measurement of atmospheric pressure. It consists of a narrow **vertically mounted** glass tube, approximately 90 cm in length, enclosed within a brass tube for protection. The glass tube is filled with mercury, and the lower end is immersed in a cistern containing mercury.

Key components include

- **Glass tube:** Contains the mercury column and is mounted vertically.
- **Brass tube:** Protects the glass tube and has a slit to view the mercury level.
- **Cistern:** A reservoir at the bottom containing mercury (P). The bottom is flexible (leather) and can be adjusted.
- **Ivory pointer:** Located inside the cistern, its tip marks the zero point of the scale (F)
- **Main scale:** A graduated scale in millimeters attached to the brass tube (M)
- **Vernier scale:** A sliding scale for precise measurements (V)
- **Adjusting screw:** Used to raise or lower the mercury level in the cistern to align with the ivory pointer.(S)

- **To take a reading**
  - Use the adjusting screw to make the mercury level in the cistern just touch the tip of the ivory pointer.
  - Observe the top of the mercury column in the glass tube.
  - Adjust the vernier scale so its edge is tangent to the meniscus (the curved surface of the mercury).
  - Read the main scale and the vernier scale to get the atmospheric pressure.

The height of the mercury column from the top of the pointer to the level of mercury in the glass tube denotes the atmospheric pressure in Cm of Hg.

- **Common units to express atmospheric pressure include**
  - **Pascals (Pa) or Hectopascals (hPa):** The standard SI unit.
  - **Millibars (mb):** Often used in meteorology (1 hPa = 1 mb).
  - **Inches of mercury (inHg) or millimeters of mercury (mmHg):** Historically used in barometers.11
  - **Atmospheres (atm):** Roughly the average pressure at sea level.

The average atmospheric pressure at sea level is about **1013.25 hPa** (or mb), which is also equal to **1 atm**.

Let's define pascal, bar, and millibar

- **Pascal (Pa):** The pascal is the SI unit of pressure. It is defined as one newton per square meter (N/m^2). In base SI units, this is equivalent to one kilogram per meter per second squared ($kg. m^{-1.} s^{-2}$). The pascal is a relatively small unit, so kilopascals (kPa) are often used.
- **Bar (bar):** The bar is a metric unit of pressure defined as exactly 100,000 pascals (105 Pa). It is slightly less than the standard atmospheric pressure.
- **Millibar (mbar):** The millibar is a unit of pressure equal to one-thousandth of a bar (10^{-3} bar). Therefore, 1 mbar=100 Pa. Millibars are commonly used in meteorology to measure atmospheric pressure.

The **newton (N)** is the standard unit of force in the International System of Units (SI).

One newton is defined as the force required to accelerate a mass of one kilogram (1 kg) at a rate of one meter per second squared ($1 m/s^2$).

Mathematically, this is expressed as: $1 N=1 kg. m/s^2$

In simpler terms, if you apply a force of one newton to an object with a mass of one kilogram, the object will accelerate at one meter per second every second.

The newton is named after Sir Isaac Newton in recognition of his work on classical mechanics, particularly his laws of motion.

8.2. Conversion of Atmospheric Pressure in Different Units

We know that 1 atm=76 Cm of Hg.

From the table given below, we can note about about the relationship between Pascals and Newtons:

Unit	**Abbreviation**	**Equivalent to 1 atm (approximately)**	**Notes**
Pascal	Pa	101325 Pa	$1 Pa=1 N/m^2$
Kilopascal	kPa	101.325 kPa	
Bar	bar	1.01325 bar	
Millibar	mbar	1013.25 mbar	
Torr	Torr	760 Torr	
Millimeters of mercury	mmHg	760 mmHg	
Centimeters of mercury	cmHg	76 cmHg	
Pounds per square inch	psi	14.696 psi	
Inches of mercury	inHg	29.92 inHg	

8.3 Gas Laws

Gas laws describe the behavior of gases under various conditions of pressure, volume, and temperature. There are several key laws:

- **Boyle's Law**: At constant temperature, the volume of a gas is inversely proportional to the pressure ($P_1V_1 = P_2V_2$).
- **Charles's Law**: At constant pressure, the volume of a gas is directly proportional to the temperature ($V_1/T_1 = V_2/T_2$).
- **Avogadro's Law**: At constant temperature and pressure, the volume of a gas is directly proportional to the number of moles ($V_1/n_1 = V_2/n_2$).
- **Ideal Gas Law**: Combines the above laws into $PV = nRT$, where P is pressure, V is volume, n is the number of moles, R is the gas constant, and T is temperature.

These laws help predict gas behavior in various situations **as explained below.**

8.3.1 Boyle's Law

- **Original**: $P_1V_1 = P_2V_2$
- **Explanation**: When temperature is constant, pressure and volume are inversely related. If pressure increases, volume decreases, and vice versa.

8.3.2 Charles's Law

- **Original:** $V_1/T_1 = V_2/T_2$
- **Explanation**: When pressure is constant, volume and temperature are directly related. If temperature increases, volume increases, and vice versa.

8.3.3 Avogadro's Law

- **Original**: $V_1/n_1 = V_2/n_2$
- **Explanation**: When temperature and pressure are constant, volume and number of moles are directly related. If the number of moles increases, volume increases.

8.3.4 Ideal Gas Law

- **Original**: $PV = nRT$
- **Explanation**: This law combines the above relationships, showing how pressure, volume, number of moles, and temperature are all connected.

8.4 Horizontal Variation of Atmospheric Pressure

The **horizontal variation of atmospheric pressure** refers to how atmospheric pressure changes across a horizontal plane at a given altitude. It's crucial for understanding and predicting weather patterns because these pressure differences drive the movement of air (wind). Here are some key aspects of the horizontal variation of atmospheric pressure:

- **Isobars:** As you mentioned, lines connecting points of equal atmospheric pressure on a map are called isobars.The pattern of isobars reveals the horizontal pressure distribution.
- **Pressure Gradient:** The rate of change of pressure with horizontal distance is called the pressure gradient.A steep pressure gradient (isobars are close together) indicates a rapid change in pressure over a short distance, which results in stronger winds. A gentle pressure gradient (isobars are far apart) means a gradual change in pressure and lighter winds.
- **Pressure Systems:** Horizontal pressure variations lead to the formation of high-pressure systems (anticyclones) and low-pressure systems (cyclones or depressions).
    - **High-pressure systems** are characterized by higher pressure in the center compared to the surroundings. In the Northern Hemisphere, air tends to flow outward and clockwise around high-pressure centers. They are often associated with stable weather.
    - **Low-pressure systems** have lower pressure in the center. In the Northern Hemisphere, air tends to flow inward and counterclockwise around low-pressure centers. They are often associated with rising air, cloud formation, and precipitation.
- **Factors Influencing Horizontal Variation:** Several factors contribute to the horizontal variation of atmospheric pressure:
    - **Temperature:** Warm air is less dense and tends to rise, leading to lower pressure. Conversely, cold air is denser and tends to sink, resulting in higher pressure. Uneven heating of the Earth's surface creates temperature differences and thus pressure variations.
    - **Earth's Rotation (Coriolis Effect):** The Earth's rotation causes a deflection of moving air (and water) to the right in the Northern Hemisphere and to the left in the Southern Hemisphere. This Coriolis effect plays a significant role in the circulation of air around pressure systems.

- **Presence of Water Vapor:** Moist air is generally lighter than dry air because water vapor molecules have less mass than the average mass of the molecules in dry air. Thus, areas with higher humidity can sometimes have slightly lower pressure.
- **Dynamic Factors:** The movement of air masses and upper-level atmospheric conditions also contribute to horizontal pressure variations.

The horizontal variation of atmospheric pressure, as depicted by isobars on weather maps, is a fundamental tool for meteorologists to understand the current weather situation and to forecast future weather conditions by analyzing the movement and evolution of pressure systems.

8.4.1 Isobaric Charts

An **isobaric chart** (also known as a weather map showing isobars) is a map that displays isobars. These lines connect locations with the same mean sea-level atmospheric pressure. By observing the patterns of isobars, meteorologists can identify areas of high and low pressure and the pressure gradient (the rate of change of pressure over distance).

- **Uses of Isobaric Charts**

Isobaric charts are extremely useful in weather forecasting and analysis because atmospheric pressure is closely related to weather patterns (Fig. 8.3). Here are some of their key uses:

- **Identifying Pressure Systems:** Isobars help to visually locate high-pressure systems (anticyclones) and low-pressure systems (cyclones or depressions). High-pressure areas are often associated with stable, fair weather, while low-pressure areas are often associated with unsettled, cloudy, and rainy or stormy weather. Closed isobars forming roughly circular or oval shapes usually indicate the centers of these systems.
- **Determining Wind Speed:** The spacing between isobars indicates the pressure gradient. When isobars are close together, it means there is a steep pressure gradient, which results in stronger winds. Widely spaced isobars indicate a weak pressure gradient and lighter winds.
- **Inferring Wind Direction:** In the Northern Hemisphere, the wind tends to blow roughly parallel to the isobars, with lower pressure to the left of the direction of the wind. In the Southern Hemisphere, it's the opposite. This relationship, along with Buys Ballot's Law, helps meteorologists estimate wind direction.

- **Locating Fronts:** Fronts (boundaries between different air masses) are often found in troughs of low pressure, which can be identified by the patterns of isobars.Isobars can show kinks or bends where fronts are located.
- **Forecasting Weather:** By analyzing the movement and changes in isobaric patterns over time, meteorologists can forecast the movement of pressure systems and associated weather conditions.

In summary, isobaric charts provide a crucial visual representation of atmospheric pressure, which is a fundamental element in understanding and predicting weather.

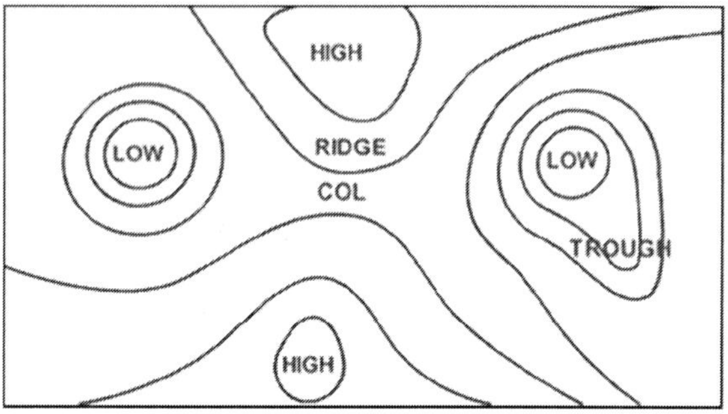

Fig. 8.3: Isobaric charts

8.4.2 Variation Pressure with Increase in Height

The variation of atmospheric pressure with an increase in height is a fundamental concept in meteorology and atmospheric science. Here's a detailed explanation:

• The Basic Principle: Gravity and the Weight of Air

Atmospheric pressure at any point is essentially the weight of the air column above that point pressing down on a unit area. Closer to the Earth's surface, there is a greater mass of air above, hence the pressure is higher. As you move to higher altitudes, the amount of air above you decreases, and consequently, the atmospheric pressure decreases.

• The Rate of Decrease: Not Linear

While pressure always decreases with height, the rate of this decrease is not linear. The air near the surface is compressed by the weight of the air above it, making it denser. As altitude increases, the air becomes less dense. This means that for the same increase in height at lower altitudes, you'll experience a larger drop in pressure compared to the same height increase at higher altitudes.

- Atmospheric pressure decreases with increasing height because there is less air above exerting pressure.
- The decrease is not linear; it's more rapid at lower altitudes where the air is denser.
- In an idealized scenario with constant temperature, the pressure decreases exponentially with height.
- In the real atmosphere, temperature variations with height (like the lapse rate in the troposphere) modify this relationship.

• Approximate Rate Near the Surface

Near the Earth's surface, for small increases in altitude, the pressure decreases roughly by about 1 hectopascal (hPa) for every 8 meters (or about 1 millibar for every 27 feet). However, this is just a rule of thumb and varies with temperature and location.

8.4.3 Pressure Gradient

The **pressure gradient** is the rate of change of atmospheric pressure over a horizontal distance. It essentially describes how quickly the pressure changes as you move from one location to another.2Mathematically, it can be represented as the change in pressure (divided by the change in distance .

$$\text{Pressure Gradient} = \frac{\text{Difference in atmospheric pressure between two points}}{\text{Distance between the two points}}$$

The pressure gradient is a **vector quantity**, meaning it has both magnitude (the rate of change) and direction (from higher to lower pressure).

8.4.4 Significance of the Pressure Gradient

The pressure gradient is of paramount significance in meteorology because it is the **primary driving force of wind**. Here's why:

- **Causes Air to Move:** Air naturally moves from areas of high pressure to areas of low pressure in an attempt to equalize the pressure difference. The pressure gradient force (PGF) is the force that arises due to this pressure difference, and it acts to accelerate air in the direction from high to low pressure.
- **Determines Wind Speed:** The magnitude (steepness) of the pressure gradient directly influences the speed of the wind.

   - A **strong pressure gradient** (isobars are close together on a weather map) indicates a rapid change in pressure over a short distance. This results in a strong pressure gradient force and thus **stronger winds.**
   - A **weak pressure gradient** (isobars are far apart) indicates a gradual change in pressure, leading to a weak pressure gradient force and **lighter winds**.
- **Direction of the Pressure Gradient Force:** The pressure gradient force always acts perpendicular to the isobars, pointing from the region of higher pressure towards the region of lower pressure.This is the initial direction of the air movement before other forces like the Coriolis force and friction come into play.
- **Formation of Weather Systems:** Horizontal variations in pressure, and thus pressure gradients, are fundamental to the formation and evolution of weather systems.
   - **Strong pressure gradients** are often associated with intense weather phenomena like cyclones, hurricanes, and strong fronts. The rapid change in pressure leads to strong winds and can contribute to rising air and precipitation.
   - **Weak pressure gradients** are typically associated with more stable weather conditions.

8.4.5 In relation to Isobaric Charts

Isobaric charts are essential for visualizing pressure gradients. As mentioned earlier:

- **Closely spaced isobars** on an isobaric chart indicate a **strong pressure gradient**.
- **Widely spaced isobars** indicate a **weak pressure gradient.**

By examining the spacing of isobars, meteorologists can quickly identify areas where strong or weak winds are likely to occur. The direction perpendicular to the isobars (pointing towards lower pressure) gives the initial direction of the pressure gradient force and the resulting wind.

In summary, the pressure gradient is a fundamental concept in understanding what drives the wind and how atmospheric pressure patterns relate to weather conditions. Its visualization on isobaric charts is a crucial tool for weather analysis and forecasting.

8.5 Wind

Wind is simply the movement of air. This movement is primarily caused by differences in air pressure, which are usually due to uneven heating of the Earth's surface by the sun.Air tends to move from areas of high pressure to areas of low pressure.

8.5.1 Wind Speed

Wind speed refers to how fast the air is moving. It is usually measured in units like kilometers per hour (km/h), miles per hour (mph), knots (nautical miles per hour), or meters per second (m/s). The strength of the wind is directly related to the pressure gradient; a greater pressure difference over a shorter distance results in higher wind speeds.

8.5.2 Wind Direction

Wind direction is the direction *from which* the wind is blowing. For example:

- A **west wind** blows from the west towards the east.
- A **north wind** blows from the north towards the south.

Wind direction is often reported using compass points (N, S, E, W, NE, SW, etc.) or in degrees from North (e.g., 0° or 360° is North, 90° is East, 180° is South, 270° is West).

8.5.3 Meanings of Windward Direction and Leeward Direction

These terms describe a position relative to the direction of the wind (Fig. 8.4):

- **Windward Direction:** The **windward** side or direction is the side that is facing *into* the wind, from where the wind is blowing.If you are standing facing the wind, you are on the windward side of yourself.
- **Leeward Direction:** The **leeward** side or direction is the side that is sheltered *from* the wind, in the direction that the wind is blowing towards. It's the downwind side. If the wind is at your back, the area in front of you is in the leeward direction relative to you.

Imagine you are standing on a beach with the wind blowing from the sea towards the land (a sea breeze).

- The side of you facing the sea (from where the wind is coming) is the **windward** side.
- The side of you facing the land (where the wind is going) is the **leeward** side.

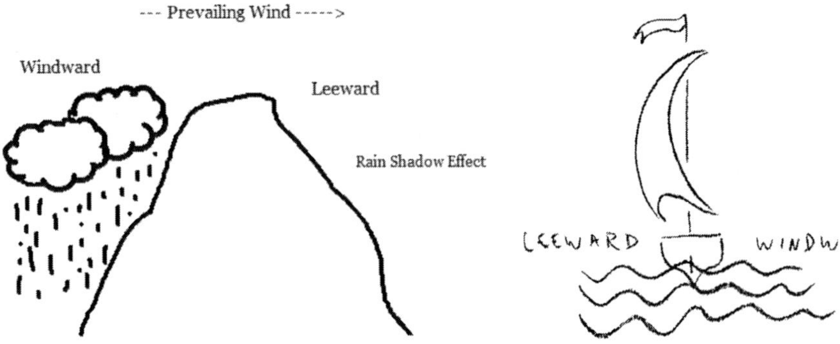

Fig. 8.4: Windward and leeward dirctions

These terms are also commonly used in other contexts, such as:

- **Sailing:** The windward side of a boat is the side facing the wind, and the leeward side is the opposite.
- **Geography:** The windward side of a mountain range receives more precipitation because the rising air cools and condenses.The leeward side often experiences a "rain shadow" and is drier.

8.6 Buys Ballots Law

Buys Ballot's Law is a principle in meteorology that describes the relationship between the direction of the wind and the distribution of atmospheric pressure. It can be stated as follows:

- **In the Northern Hemisphere:** If you stand with your back to the wind, the area of lower atmospheric pressure will be to your left, and the area of higher pressure will be to your right.
- **In the Southern Hemisphere:** This is reversed. If you stand with your back to the wind, the area of lower atmospheric pressure will be to your right, and the area of higher pressure will be to your left.

In simpler terms: Low pressure is on your left in the Northern Hemisphere and on your right in the Southern Hemisphere when the wind is at your back.

This law is a consequence of the **Coriolis effect**, which is caused by the Earth's rotation. The Coriolis effect deflects moving objects (including air) to the right in the Northern Hemisphere and to the left in the Southern Hemisphere.

When air moves from high to low pressure, it doesn't travel directly along the pressure gradient. Instead, it is deflected by the Coriolis force. This deflection results in the wind circulating around pressure systems:

- Around a **low-pressure system** in the Northern Hemisphere, the air flows counterclockwise and inwards.
- Around a **high-pressure system** in the Northern Hemisphere, the air flows clockwise and outwards.
- The directions are reversed in the Southern Hemisphere.

8.6.1 Significance of Buys Ballot's Law

- **Locating Pressure Systems:** Sailors and meteorologists can use this law to get a rough idea of the location of high and low-pressure systems simply by noting the wind direction. This can be particularly useful at sea where direct pressure readings across a wide area are not available. For instance, a mariner in the Northern Hemisphere experiencing a wind that is veering (changing clockwise) knows that a high-pressure area might be passing to their right, or a low-pressure area to their left is moving away. A backing wind (changing counterclockwise) suggests the approach of a low-pressure system to their left.
- **Navigating Storms:** The law has implications for avoiding the most dangerous parts of storms. For example, in the Northern Hemisphere, the right front quadrant of a hurricane is often the most dangerous because the storm's rotational winds are compounded by its forward motion. Knowing Buys Ballot's Law can help in steering away from this quadrant.

Buys Ballot's Law is an approximation. Near the Earth's surface, friction with the ground can alter the wind direction, causing it to cross the isobars at an angle rather than being parallel to them. However, it's a useful rule of thumb, especially in the free atmosphere above the influence of surface friction.

8.7 Global Pressure and Wind Systems

Global pressure belts and wind systems are large-scale patterns of atmospheric circulation that shape our climate and weather. Let's break it down:

8.7.1 Global Pressure Belts

The Earth's atmosphere is divided into several pressure belts, which are areas of high or low pressure that encircle the globe (Fig. 8.5). These belts are created by the uneven heating of the Earth's surface by the sun.

- **Equatorial Low-Pressure Belt**: Near the equator, the sun's rays strike the Earth directly, heating the air and causing it to rise. This creates a belt of low pressure near the equator.

- **Subtropical High-Pressure Belts**: At around 30°N and 30°S latitude, the air cools and sinks, forming high-pressure belts. These belts are also known as the horse latitudes.
- **Subpolar Low-Pressure Belts**: At around 60°N and 60°S latitude, the air cools and rises, creating low-pressure belts.
- **Polar High-Pressure Belts**: At the poles, the air is cold and dense, forming high-pressure belts.

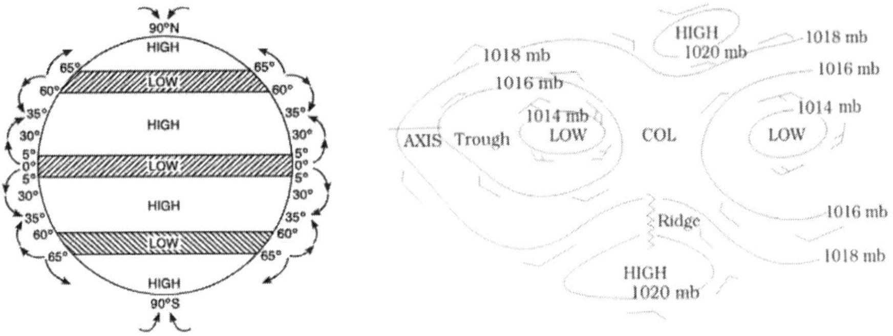

Fig. 8.5: Atmospheric pressure belts and wind systems

8.7.2 Global Wind Systems

Wind is the movement of air from high-pressure areas to low-pressure areas. The global wind systems (Fig. 8.6) are driven by the pressure differences between the pressure belts.

- **Trade Winds**: Blow from the subtropical high-pressure belts towards the equatorial low-pressure belt. In the Northern Hemisphere, trade winds blow from the northeast, while in the Southern Hemisphere, they blow from the southeast.
- **Westerlies**: Blow from the subtropical high-pressure belts towards the subpolar low-pressure belts. In the mid-latitudes, westerlies blow from the west towards the east.
- **Polar Easterlies**: Blow from the polar high-pressure belts towards the subpolar low-pressure belts. These winds are cold and dry.

The global pressure belts and wind systems work together to create a circulation pattern in the atmosphere. Air rises at the equator, creating a low-pressure belt, and sinks at the subtropics, creating a high-pressure belt. This circulation pattern drives the trade winds, westerlies, and polar easterlies.

Understanding global pressure belts and wind systems is crucial for predicting weather patterns, climate, and ocean currents. These patterns play a significant role in shaping regional climates, influencing precipitation, and impacting global weather phenomena like hurricanes and monsoons.

In summary, global pressure belts and wind systems are large-scale patterns of atmospheric circulation that drive the Earth's climate and weather. The pressure belts are created by uneven heating of the Earth's surface, and the wind systems are driven by pressure differences between the belts.

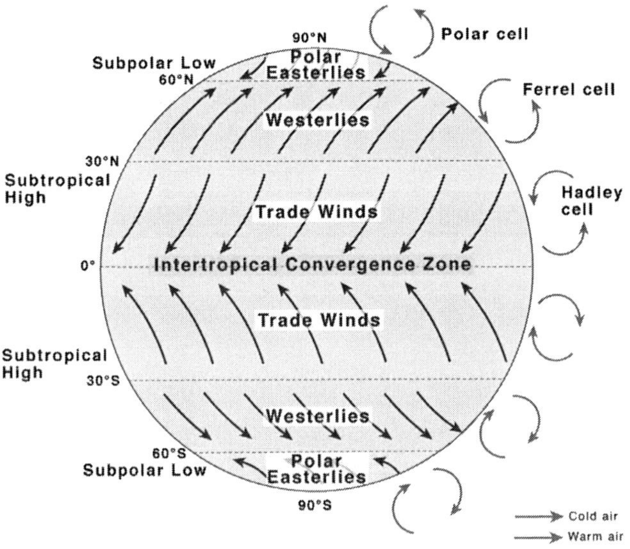

Fig. 8.6: Global wind systems

8.8 Daily and Seasonal Variation of Wind Speed

Daily and seasonal variations in wind speed are significant factors in weather patterns, climate, and renewable energy production. Here's a breakdown of each:

8.8.1 Daily (Diurnal) Variations

Daily variations in wind speed are primarily driven by the **heating and cooling of the Earth's surface**throughout the day.

- **Daytime:** Solar radiation heats the ground, which in turn heats the air near the surface. This warmer air rises (convection), leading to increased vertical mixing in the atmosphere. This mixing can bring down higher-speed winds from aloft, resulting in stronger surface winds during the day, typically peaking in the afternoon.

- **Nighttime:** As the sun sets, the ground cools, and the air near the surface also cools and becomes more stable. This stability reduces vertical mixing, and surface friction becomes a more dominant factor. Consequently, surface wind speeds often decrease at night. However, in some coastal regions, land breezes (winds blowing from land to sea at night) can develop. Mountain and valley breezes also exhibit diurnal reversals due to differential heating and cooling of slopes and valleys.

8.8.2 Seasonal Variations

Seasonal variations in wind speed are influenced by **larger-scale weather systems and temperature differences** that change throughout the year.7

- **Mid-latitudes:** Many mid-latitude regions experience stronger winds during the late fall, winter, and spring due to more intense and frequent storm systems (like mid-latitude cyclones) driven by larger temperature gradients between the poles and the equator. Summer often sees weaker winds as these storm systems are generally less intense and less frequent.
- **Coastal Regions:** Seasonal differences in the temperature contrast between land and sea can enhance or weaken coastal breezes, affecting overall wind speeds.
- **Monsoon Regions:** Areas with monsoon climates can see significant seasonal shifts in wind speed and direction.9 For example, the Indian subcontinent experiences strong southwesterly winds during the summer monsoon.

8.8.3 Factors Influencing Wind Speed Variations

Several factors contribute to both daily and seasonal variations in wind speed:

- **Pressure Gradients:** The ratio of difference in atmospheric pressure between two points to the distance between the two points is called pressure gradient. Differences in air pressure cause air to move from high to low pressure. The greater the pressure gradient , the stronger the wind speed.These pressure gradients change daily and seasonally.
- **Temperature Differences:** Uneven heating of the Earth's surface (both daily and seasonally) leads to pressure differences, driving wind.
- **Coriolis Effect:** The Earth's rotation deflects moving air (to the right in the Northern Hemisphere and to the left in the Southern Hemisphere). This effect influences the direction16 of the wind more significantly at larger scales but also plays a role in the overall wind patterns that vary seasonally.

- **Friction:** The Earth's surface slows down the wind, especially near the ground.This effect is more pronounced during the day when vertical mixing is weaker near the surface at night. The type of surface (e.g., smooth water vs. rough terrain) also affects friction.
- **Local Geography:** Features like mountains, valleys, and coastlines can create localized wind patterns and modify regional wind speeds both daily and seasonally.

Understanding these daily and seasonal variations is crucial for various applications, including weather forecasting, climate modeling, agriculture, and optimizing the efficiency of wind energy generation.

8.9 Different Types of Winds

• Planetary or Permanent Winds (Global Winds)

These are large-scale wind systems that circulate across the globe. They are relatively constant in direction and are primarily driven by the uneven heating of the Earth and the Coriolis effect (due to Earth's rotation). This uneven heating creates zones of high and low pressure, leading to consistent air movement.

8.9.1 Trade Winds

- **Explanation:** Near the equator, the sun's rays are most direct, heating the air and causing it to rise (creating the Intertropical Convergence Zone - ITCZ, a low-pressure belt). Higher up, this air moves poleward, cools, and descends around 30 degrees latitude north and south (subtropical high-pressure belts). From these high-pressure belts, air flows back towards the equator. Due to the Coriolis effect, these winds are deflected: to the right in the Northern Hemisphere, becoming **northeasterly trade winds**, and to the left in the Southern Hemisphere, becoming **southeasterly trade winds**.
- **Significance:** Historically crucial for sailing ships traveling across the oceans. They also influence weather patterns in tropical regions.

8.9.2 Westerlies

- **Explanation:** Originating from the subtropical high-pressure belts (around 30 degrees latitude), some of the descending air flows towards the poles. Again, the Coriolis effect deflects these winds: to the right in the Northern Hemisphere, resulting in generally **southwesterly westerlies**, and to the left in the Southern Hemisphere, resulting in generally **northwesterly westerlies**.

- **Significance:** These winds dominate the mid-latitudes and are responsible for much of the weather movement across continents in these zones. They are particularly strong in the Southern Hemisphere because there is less land to obstruct their flow, leading to the "Roaring Forties" (40°S), "Furious Fifties" (50°S), and "Shrieking Sixties" (60°S).
- **Polar Easterlies**
- **Explanation:** At the poles, cold, dense air sinks, creating high-pressure areas (polar highs). This cold air then flows away from the poles towards the lower latitudes (around 60 degrees latitude, the subpolar low-pressure belts). The Coriolis effect deflects these winds: to the right in the Northern Hemisphere, creating **northeasterly polar easterlies**, and to the left in the Southern Hemisphere, creating **southeasterly polar easterlies**.
- **Significance:** These are generally weak and irregular winds compared to the trades and westerlies.

• Periodic or Seasonal Winds

These wind patterns change their direction with the seasons due to variations in temperature and pressure that occur over longer periods (like months).

8.9.3 Monsoon Winds

- **Explanation:** The most well-known monsoon system affects South and Southeast Asia. During summer, the landmass heats up significantly more than the surrounding oceans, creating a low-pressure area over land. This draws in moist air from the relatively high-pressure over the Indian Ocean, resulting in the **summer monsoon**, characterized by heavy rainfall. In winter, the land cools down more quickly, creating a high-pressure area, and the winds reverse, blowing dry air from the land towards the ocean – the **winter monsoon**.
- **Significance:** Crucial for agriculture in many regions, bringing much-needed rainfall. However, they can also cause devastating floods.

8.9.4 Land and Sea Breezes (Diurnal Winds)

- **Explanation:** These occur daily near coastlines. During the day, land heats up faster than water. The warmer air over land rises, creating a low-pressure area, and cooler air from over the sea moves in to replace it, resulting in a **sea breeze** (wind from sea to land). At night, the land cools faster than the sea. Now, the air over the sea is warmer and rises, and cooler air from the land moves towards the sea, creating a **land breeze** (wind from land to sea).

- **Significance:** Affect local weather and can be important for coastal communities.

8.9.5 Mountain and Valley Breezes (Diurnal Winds in Mountainous Regions)

- **Explanation:** Similar to land and sea breezes, but driven by the differential heating and cooling of mountain slopes and valleys. During the day, the mountain slopes heat up more quickly than the valley floor. The warmer air on the slopes rises, creating a **valley breeze** (wind blowing uphill). At night, the slopes cool down faster, and the denser, cooler air descends into the valley, creating a **mountain breeze** (wind blowing downhill).
- **Significance:** Influence local temperatures and can play a role in the formation of clouds and precipitation in mountainous areas.

8.9.6 Local Winds

These are winds that affect relatively small areas and are often caused by specific local geographical features or conditions.

- **Hot Local Winds:** (Descriptions remain similar, highlighting the region and characteristics)
  - **Loo:** Hot, dry, and dusty winds blowing across the northern plains of India and Pakistan during the summer months.
  - **Chinook:** A warm, dry, downslope wind on the eastern side of the Rocky Mountains in North America. It can cause rapid temperature increases and snowmelt.
  - **Foehn:** A warm, dry wind that descends the leeward (downwind) side of a mountain range, particularly in the Alps. The air loses moisture on the windward side and warms adiabatically as it descends.
  - **Sirocco:** A hot, often humid wind from the Sahara Desert that blows northward across the Mediterranean Sea, affecting Southern Europe. It can carry dust and sand.
- **Cold Local Winds:** (Descriptions remain similar, highlighting the region and characteristics)
  - **Mistral:** A strong, cold, and typically dry northwesterly wind that blows from southern France into the Gulf of Lion in the Mediterranean.
  - **Bora:** A cold, often gusty, and dry northeasterly wind that blows down from the highlands bordering the Adriatic Sea.

- **Blizzard:** A severe winter storm characterized by strong winds (at least 35 mph), heavy snowfall, and low visibility (less than a quarter mile) for an extended period.

8.10 Wind Measurements

The wind speeds are measured using an anemometer and wind direction is measured using a wind vane.

8.10.1 Cup Anemometer

- **Construction :** Similar to the general cup anemometer, a manual observation type often features:
  - **Cups:** Typically three hemispherical cups.
  - **Horizontal Arms:** Holding the cups, radiating from a central shaft.
  - **Vertical Shaft:** That rotates with the cups.
  - **Counter Mechanism:** This is the key difference. Instead of an electronic transducer, a manual observation anemometer is linked to a mechanical counter. This counter records the total "runs" of the wind passing the instrument. Each count on the counter usually corresponds to a specific distance of air movement (e.g., 100 meters).

Based on research, IMD manufactures a **cup counter anemometer sensor**, which aligns with this description.

- **Working Principle**
  - Wind causes the cups to rotate.
  - This rotation is mechanically linked to the counter mechanism.
  - The counter increments with the rotation, effectively measuring the total distance of the wind that has passed through the anemometer over a period.

To determine the wind speed at a specific time of observation:

- An observer takes an initial reading of the counter.
- After a measured interval of time (e.g., 3 minutes), the observer takes a second reading.
- The difference between the two readings gives the "wind run" (total distance of wind).
- The average wind speed during that interval is calculated by dividing the wind run by the time interval.

For example, if the counter reading changes by 2.2 units in 3 minutes, and each unit represents 100 meters of wind run:

- Wind run = $2.2 \times 100 = 220$ meters
- Time interval = 3 minutes = $3/60 = 0.05$ hours
- Average wind speed = 3 minutes220 meters =180 seconds220 meters ≈ 1.22 m/s
- Or, in km/hour: 0.05 hours0.22 km =4.4 km/hour

IMD's standard measurement practice involves taking two successive readings at a 3-minute interval to determine the instantaneous wind speed at the time of observation (typically 0830 and 1730 IST).

- **Installation :** Installation is similar to automated versions, focusing on:
  - **Siting:** Open area, away from obstructions, and at a representative height
  - **Mounting:** Securely fixed to a mast to prevent movement.
  - **Free Rotation:** Ensure the cups rotate freely without any obstruction.
- **Measurement**
  - At the time of observation, note the reading on the mechanical counter.
  - Start a stopwatch and after a specific time interval (e.g., 3 minutes), note the new counter reading.
  - Calculate the difference in the readings (wind run).
  - Convert the wind run to distance (if the counter units aren't already in meters or kilometers based on the anemometer's calibration).
  - Divide the distance by the time interval to get the average wind speed during that period.

IMD's manual on hydro-meteorology describes this process, where the difference in counter readings over 3 minutes (where 1 count equals 100 m) is multiplied by 20 to get the mean wind speed in km/hour.

8.10.2 Wind Vane

Let's discuss the manual observation of wind direction using a wind vane, potentially one adhering to IMD specifications.

- **Construction:** A typical wind vane for manual observation consists of the following key components:
  - **Pointer (or Arrowhead):** A streamlined shape, often an arrow or a similar pointed design, that aligns itself with the wind. It's usually

made of lightweight but rigid material like sheet metal (aluminum) or durable plastic.

- **Tail:** A larger, often flat or finned surface at the opposite end of the pointer. The pressure difference caused by the wind on the tail forces the pointer to swing and align with the direction from which the wind is blowing.
- **Vertical Shaft:** The pointer and tail assembly is mounted on a vertical shaft that allows it to rotate freely in the horizontal plane.
- **Bearings:** Low-friction bearings are crucial for the vane to respond sensitively to changes in wind direction.
- **Directional Indicators:** Below the rotating vane, a fixed compass rose or a set of cardinal (N, S, E, W) and sometimes intercardinal (NE, SE, SW, NW) markings are present. These help the observer visually determine the wind direction. For more precise readings, finer divisions (e.g., to 16 points of the compass) might be marked.
- **Mounting Base:** A structure to securely fix the wind vane assembly to a mast or pole.

IMD often uses wind vanes in conjunction with cup anemometers. Their wind vanes are typically robust for continuous outdoor use. Some of their systems utilize potentiometric wind vanes for automated readings, but for manual observation, the principle of a free-rotating vane with directional markings is fundamental.

- **Installation:** Proper installation is crucial for accurate wind direction readings:
  - **Siting:** Similar to anemometers, the wind vane should be placed in an open area, away from obstructions that could create localized wind eddies or block the true wind flow. It should ideally be mounted above the height of nearby obstacles.
  - **Mounting:** The vane assembly needs to be securely attached to a vertical pole or mast. Ensure it is level and the vertical shaft is truly vertical to allow free horizontal rotation.
  - **Orientation:** This is the most critical step for a wind vane used for absolute direction. The fixed directional indicators (N, S, E, W) *must* be accurately aligned with the true cardinal directions. This is typically done using a magnetic compass, accounting for the local magnetic declination. Once aligned with magnetic north, the markings are adjusted to represent true north.

- **Method of Observation:** Manual observation of wind direction using a wind vane is a visual process:
    - **Observation Point:** The observer needs a clear line of sight to the wind vane and its directional markings.
    - **Alignment:** Wait for the wind vane to settle and align itself consistently with the prevailing wind. Due to gusts and turbulence, it might fluctuate slightly, so the observer needs to note the average direction indicated by the pointer.
    - **Reading the Direction:** Observe which of the fixed directional markings the pointer (arrowhead) is currently pointing towards. The wind direction is reported as the direction *from which* the wind is blowing.
        - If the arrow is pointing towards 'N', the wind direction is North.
        - If it's between 'N' and 'E', it would be reported as Northeast (NE), and so on.
        - If finer markings are present, the direction can be estimated more precisely (e.g., North-Northwest).

IMD practices for manual weather observations would involve trained personnel visually assessing the wind vane and recording the direction using standard meteorological terminology (e.g., N, NNE, NE, ENE, E, etc., often to 16 points of the compass). These observations are usually made at specific synoptic hours.

8.11 Construction of Wind Rose

A **wind rose** is a graphical tool used by meteorologists and others to illustrate the distribution of wind direction and wind speed at a particular location over a period of time (Fig. 8.7). It essentially summarizes how often the wind blows from different directions and at what speeds.

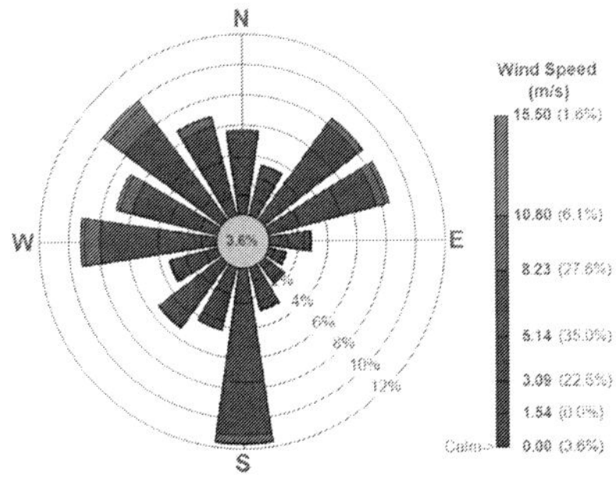

Fig. 8.7: Wind rose

Construction: A wind rose is typically constructed using historical wind data collected at a specific location. The process involves these general steps:

- **Data Collection:** Wind speed and direction data are gathered over a significant period (e.g., months, a year, or multiple years) from a reliable source like a weather station.These data points usually include both the direction the wind is blowing *from* and its speed.
- **Data Processing:** The collected data is then analyzed to determine:
  - The frequency of winds from different directional sectors (e.g., North, North-Northeast, Northeast, etc.). Typically, 8 or 16 directional sectors are used.
  - The distribution of wind speeds within each of these directional sectors. Wind speeds are often categorized into ranges (e.g., 0-5 km/h, 5-10 km/h, etc.).
  - The frequency of calm conditions (very low or no wind).
- **Diagram Creation:** The wind rose is then plotted, typically as a circular diagram with the following features:
- **Directional Spokes:** Lines (spokes) radiate from a central point, each representing a different wind direction. The orientation of these spokes corresponds to the compass directions.
- **Frequency:** The length of each spoke indicates how frequently the wind blows from that particular direction. Longer spokes mean more frequent winds from that direction. Concentric circles emanating from the center can also represent frequency, allowing for a quantitative assessment.
- **Wind Speed:** Often, the spokes are divided into segments of different colors or shading. Each segment represents a different wind speed range. The length of a colored segment along a spoke shows how frequently the wind blows from that direction *at that particular speed range.*
- **Calm Conditions:** The frequency of calm winds is often indicated by a circle or a value in the center of the wind rose.

In essence, to construct a wind rose, you need a dataset of wind speed and direction, you categorize this data by direction and speed, and then you visually represent these frequencies using a circular graph with spokes for directions and segments along the spokes for speed ranges.

Atmospheric Pressure and Winds

1. Atmospheric pressure is defined as:

a) The weight of the ocean above sea level
b) The weight of the air column above a unit area
c) The temperature of the air
d) The humidity of the atmosphere

2. Standard atmospheric pressure at sea level is:

a) 101.3 kPa
b) 1.00 atm
c) 760 mm Hg
d) All of the above

3. The instrument used to measure atmospheric pressure is called:

a) Thermometer
b) Hygrometer
c) Barometer
d) Anemometer

4. Atmospheric pressure decreases with:

a) Increasing altitude
b) Increasing temperature
c) Both a and b
d) Increasing humidity only

5. Pressure belts around the globe are created due to:

a) Uneven heating of Earth's surface
b) Earth's magnetic field
c) Ocean currents
d) Moon's gravitational pull

6. Pressure decreases by approximately how much per 100 m increase in elevation?

a) 1 mb
b) 10 mb
c) 12 mb
d) 20 mb

7. The equator is generally a zone of:

a) Low pressure
b) High pressure
c) Variable pressure
d) No pressure

8. Subtropical regions are characterized by:

a) Low rainfall and high pressure
b) High rainfall and low pressure
c) High winds and low pressure
d) High cloud cover

9. High-pressure systems are usually associated with:

a) Cloudy weather
b) Storms
c) Clear skies
d) Heavy rainfall

10. In a low-pressure system, air tends to:

a) Descend and diverge
b) Rise and converge
c) Remain stagnant
d) Sink and rotate clockwise

11. Wind is defined as the movement of air from:

a) Low to high pressure
b) High to low pressure
c) West to east only
d) Ocean to land

12. The primary force responsible for wind is:

a) Gravity
b) Friction
c) Coriolis force
d) Pressure gradient force

13. The Coriolis effect is caused by:

a) Moon's gravity
b) Earth's rotation
c) Ocean tides
d) Earth's revolution around the sun

14. In the Northern Hemisphere, Coriolis force deflects winds to the:

a) Left
b) Right
c) Center
d) Backward

15. Geostrophic winds result from a balance between:

a) Gravity and friction
b) Pressure gradient and Coriolis force
c) Temperature and pressure
d) Friction and centrifugal force

16. Trade winds blow from:

a) Equator to poles
b) Subtropics to equator
c) Poles to subtropics
d) East to west at poles

17. Westerlies are winds that blow:

a) From west to east between 30°–60° latitudes
b) From east to west near equator
c) From poles to equator
d) Over deserts only

18. Polar easterlies blow from:

a) 0°–30°
b) 30°–60°
c) 60°–90°
d) 90°–120°

19. Doldrums are found near:

a) 30° latitude
b) 0° latitude
c) 45° latitude
d) 90° latitude

20. The horse latitudes are located at:

a) 0°
b) 15°
c) 30°
d) 60°

21. A sea breeze blows from:

a) Land to sea during day
b) Sea to land during day
c) Sea to land during night
d) Land to sea during monsoon

22. Land breeze occurs when:

a) Land is cooler than sea at night
b) Sea is warmer than land during day
c) Sea is frozen
d) Sun is overhead

23. The hot, dry local wind of northern India is called:

a) Chinook
b) Mistral
c) Loo
d) Harmattan

24. The warm, dry wind descending the leeward slope of mountains is known as:

a) Cyclone
b) Chinook
c) Monsoon
d) Trade wind

25. Monsoon winds are:

a) Periodic winds
b) Local winds
c) Planetary winds
d) Cyclonic winds

26. Cyclones are centers of:

a) High pressure with outward winds
b) Low pressure with inward rotating winds
c) High temperature and pressure
d) No wind movement

27. In the Northern Hemisphere, cyclones rotate:

a) Clockwise
b) Anticlockwise
c) Do not rotate
d) Vertically

28. Anticyclones are associated with:

a) Rising air and cloud formation
b) Sinking air and clear skies
c) Strong storms
d) High rainfall

29. Cyclones in the Indian Ocean are also referred to as:

a) Hurricanes
b) Typhoons
c) Tropical cyclones
d) Tornadoes

30. The calm center of a tropical cyclone is called:

a) Core
b) Vortex
c) Eye
d) Nucleus

31. The speed of wind is measured using:

a) Barometer
b) Thermometer
c) Anemometer
d) Hygrometer

32. Wind direction is measured using:

a) Wind vane
b) Rain gauge
c) Thermograph
d) Wind barometer

33. Beaufort scale is used to:

a) Measure air temperature
b) Classify rainfall types
c) Estimate wind speed visually
d) Measure cloud cover

34.The unit of wind speed in meteorology is commonly:

a) Degrees
b) mm/hr
c) m/s or km/h
d) hPa

35. Isobars are lines connecting points of equal:

a) Rainfall
b) Wind direction
c) Atmospheric pressure
d) Temperature

36. Surface winds over water bodies influence:

a) Soil erosion only
b) Temperature of deep oceans
c) Wave generation and mixing of water
d) Evapotranspiration from crops

37. Strong winds over lakes and ponds can lead to:

a) Thermal pollution
b) Stratification
c) De-stratification or overturning
d) Cloud formation

38. Wind-induced mixing enhances:

a) Evaporation but reduces oxygenation
b) Oxygen distribution in water columns
c) Temperature stability in water
d) Sedimentation

39. Persistent wind from one direction causes:

a) Equal pressure across lakes
b) Water piling at one end of the lake
c) Decrease in wind velocity
d) Air stagnation

40. Fisheries in shallow water bodies are sensitive to:

a) Cloud height
b) Diurnal wind cycles
c) Satellite radiation
d) Altitude pressure

41. Hadley cells operate between:

a) 0°–30° latitude
b) 30°–60° latitude
c) 60°–90° latitude
d) 45°–75° latitude

42. Jet streams are:

a) Surface ocean currents
b) High-altitude, high-speed winds
c) Tornadoes
d) Local winds

43. Polar front theory explains the formation of:

a) Cyclones in equatorial regions
b) Mid-latitude depressions
c) Trade winds
d) Monsoon onset

44. The Intertropical Convergence Zone (ITCZ) is a zone of:

a) Diverging winds
b) Descending air
c) Converging trade winds and rising air
d) Permanent high pressure

45. Seasonal shifts in ITCZ influence:

a) Tidal cycles
b) Monsoon rainfall patterns
c) Evaporation rates in deserts
d) Jet stream intensity

46. Wind energy is harnessed using:

a) Solar panels
b) Anemometers
c) Wind turbines
d) Watermills

47. Wind chill factor refers to:

a) Cooling effect of wind on warm air
b) Rise in air temperature due to wind
c) Perceived drop in temperature due to wind
d) Wind causing rainfall

48. Frictional force affects winds near:

a) Troposphere top
b) Earth's surface
c) Equator only
d) Polar regions

49. Pressure gradients are steeper when isobars are:

a) Far apart
b) Equidistant
c) Close together
d) Curved

50. Which of the following best explains wind convergence at low pressure zones?

a) Winds blow outward to balance pressure
b) Winds rise vertically and cool
c) Winds spiral inward due to pressure gradient and Coriolis force
d) There is no convergence

Answer Key (s)

1	b	2	d	3	c	4	c	5	a	6	b	7	a	8	c	9	c	10	b
11	b	12	d	13	b	14	b	15	b	16	b	17	a	18	c	19	b	20	c
21	b	22	a	23	c	24	b	25	a	26	b	27	b	28	b	29	c	30	c
31	c	32	a	33	c	34	c	35	c	36	c	37	c	38	b	39	b	40	b
41	a	42	b	43	b	44	c	45	b	46	c	47	c	48	b	49	c	50	c

9

Land and Sea Breezes

As you deepen your understanding of aquatic ecosystems, it's essential to recognize that the ocean and its inhabitants are not isolated entities. They are intricately linked with the atmosphere above. This chapter focuses on a fascinating and directly relevant atmospheric phenomenon for coastal environments: **land and sea breezes** (Fig. 9.1).

These aren't just pleasant onshore or offshore flows of air. Land and sea breezes represent a daily cycle of wind patterns driven by the fundamental difference in how land and water absorb and release heat. This seemingly simple process has far-reaching consequences for the nearshore marine environment, influencing everything from the subtle shifts in water temperature and salinity to the crucial distribution of nutrients that underpin the entire marine food web.

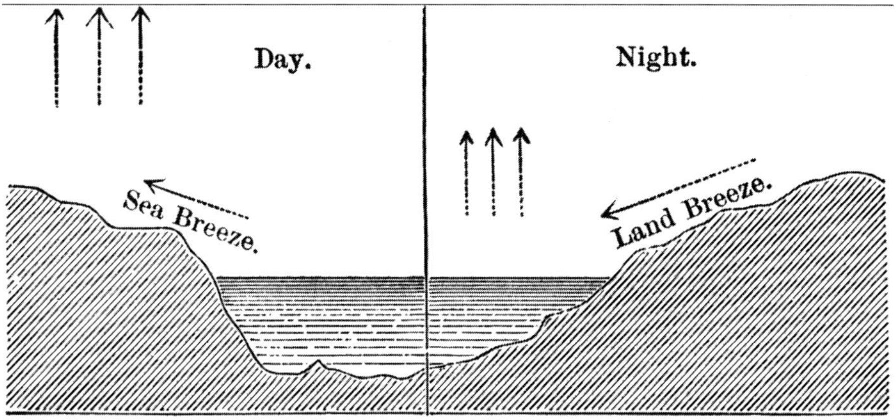

Fig. 9.1: Land and sea breezes

For those of you who will be involved in the intricate world of aquaculture, the art and science of fishing, or the vital task of managing our precious marine resources, understanding the dynamics of land and sea breezes will provide you with a powerful lens through which to interpret the daily and localized changes within coastal aquatic ecosystems. This knowledge can inform decisions about optimal fishing times, the placement of aquaculture farms, and the potential impacts of environmental changes.

In this chapter, we will delve into

- **The underlying physics:** We'll unravel the science behind why land heats and cools faster than water, and how this temperature difference initiates the movement of air.
- **The diurnal cycle:** We will trace the evolution of these breezes throughout the day and night, from the gentle onshore flow of a sea breeze to the subtle offshore drift of a land breeze.
- **The ecological significance for fisheries:** Crucially, we will explore the diverse ways these winds influence the aquatic environment that directly impact fisheries. This includes how they can affect:
    - **Water temperature stratification:** Leading to variations in habitat suitability for different species.
    - **Nutrient mixing:** Bringing vital elements to the surface and influencing primary productivity.
    - **Plankton distribution:** The base of many marine food chains.
    - **The movement and aggregation of fish:** Potentially influencing fishing success.
    - **Coastal currents:** Affecting larval dispersal and the transport of pollutants.

By the end of this chapter, you will appreciate that the daily "rhythm of the coast," dictated in part by land and sea breezes, plays a more profound role in the lives of aquatic organisms and the success of fisheries than might initially meet the eye. Let's embark on this enlightening journey together!

9.1 Mechanism of Land and Sea Breezes

Okay, let's delve into the detailed mechanism of land and sea breezes:

The occurrence of land and sea breezes is primarily due to the **differential heating and cooling rates of land and water**. Water has a higher **specific heat capacity** than land, meaning it takes more energy to change the temperature of water compared to land, and it also retains heat longer. This difference leads to temperature and consequently pressure variations near the coast over a 24-hour cycle.

Here's a breakdown of how each type of breeze forms:

9.1.1 Sea Breeze (Occurs During the Day)

- **Solar Heating:** During the day, the sun's radiation heats both the land and the sea. However, the land surface heats up much more quickly than the sea surface.

- **Air Warming Over Land:** As the land surface warms, it heats the air directly above it through **conduction**. This warmer air near the surface then rises due to **convection** (warm air is less dense).
- **Low Pressure Over Land:** The rising of warm air over the land creates a region of lower atmospheric pressure at the surface.
- **High Pressure Over Sea:** Meanwhile, the sea surface warms up much more slowly. The air above the sea remains relatively cooler and denser, resulting in a region of higher atmospheric pressure at the surface compared to the land.
- **Air Flow from Sea to Land:** Air naturally flows from areas of high pressure to areas of low pressure. Therefore, the cooler, denser air over the sea moves horizontally towards the land to replace the rising warmer air. This movement of air from the sea towards the land is what we experience as a **sea breeze**.
- **Upper Circulation:** To complete the circulation, the warmer air that rose over the land eventually cools at higher altitudes and then sinks over the sea, creating a return flow aloft from land to sea. This forms a localized circulation cell.

In summary, a sea breeze is a wind that blows from the sea towards the land during the day because the land heats up faster than the sea, creating a low-pressure area over land that draws in cooler air from the higher-pressure area over the sea.

9.1.2 Land Breeze (Occurs During the Night)

- **Cooling after Sunset:** After sunset, both the land and the sea begin to cool down. However, just as the land heated up faster, it also cools down much more quickly than the sea.
- **Air Cooling Over Land:** The land surface rapidly loses heat through radiation, cooling the air directly above it. This cooler air becomes denser and tends to sink.
- **High Pressure Over Land:** The sinking of cooler, denser air over the land leads to the development of a region of higher atmospheric pressure at the surface.
- **Low Pressure Over Sea:** The sea, having retained its heat from the day, cools down much more slowly. The air above the sea remains relatively warmer and less dense, resulting in a region of lower atmospheric pressure at the surface compared to the land.

- **Air Flow from Land to Sea:** Again, air flows from high to low pressure. Therefore, the cooler, denser air over the land moves horizontally towards the sea to replace the rising warmer air over the water. This movement of air from the land towards the sea is what we experience as a **land breeze**.
- **Upper Circulation:** Similarly to the sea breeze, a return flow completes the circulation. The warmer air that rose over the sea cools at higher altitudes and then sinks over the land, creating a flow aloft from sea to land.

In summary, a land breeze is a wind that blows from the land towards the sea during the night because the land cools down faster than the sea, creating a high-pressure area over land that forces cooler air towards the lower-pressure area over the warmer sea. Key Factors:

- **Specific Heat Capacity:** The primary driver.
- **Convection:** Vertical air movement due to temperature differences.
- **Pressure Gradient Force:** The force that moves air from high to low pressure.

9.1.3 Significance of Land and Sea Breezes in Marine Fisheries

The significance of land and sea breezes extends into the realm of marine fisheries in several important ways:

- **Influence on Water Temperature**
  - **Sea breezes (daytime):** Can bring cooler air from over the sea, potentially moderating the surface water temperatures near the coast. This can be crucial for the comfort and distribution of temperature-sensitive fish species.
  - **Land breezes (nighttime):** May bring slightly cooler land air offshore, potentially leading to minor temperature changes in nearshore waters.
- **Nutrient Mixing and Upwelling**
  - While not as strong as major upwelling events, sea breezes can sometimes contribute to localized, nearshore upwelling. As the surface water near the land moves offshore due to the returning upper flow of the sea breeze circulation, it can be replaced by slightly deeper, nutrient-rich water. This increased nutrient availability can enhance primary productivity (phytoplankton growth), which forms the base of the marine food web, ultimately benefiting fish populations.

- **Plankton Distribution**
  - The movement of surface waters associated with land and sea breezes can influence the distribution of plankton, the primary food source for many small fish and larvae. Sea breezes might push surface plankton towards the shore, while land breezes could move it slightly offshore. This can affect where smaller fish congregate and, consequently, where larger predatory fish might be found.
- **Fish Behavior and Location**
  - Some fish species may exhibit behavioral changes in response to the subtle shifts in temperature, currents, and plankton distribution caused by land and sea breezes. Fishermen with local knowledge often learn these patterns over time. For example, certain fish might move closer to shore with a sea breeze if it brings cooler water or a concentration of food.
- **Fishing Conditions and Safety**
  - **Guidance for Fishermen:** Traditional fishermen, in particular, often use the predictable cycle of land and sea breezes to their advantage:
    - **Morning:** They might sail offshore with the land breeze (blowing from land to sea) and return in the afternoon with the sea breeze (blowing from sea to land). This was especially important for sail-powered vessels.
    - **Timing of fishing:** Experienced fishermen often know that certain times of the day, influenced by the onset or change in these breezes, can be better for catching specific types of fish due to the factors mentioned above (temperature, plankton).
    - **Safety:** Understanding the daily wind patterns helps in planning fishing trips and anticipating changes in sea conditions. A sudden change or strengthening of the breeze can indicate a need to adjust fishing locations or return to shore.
- **How Fishermen are Guided**
  - **Traditional Knowledge:** Many coastal fishing communities have accumulated generations of knowledge linking wind patterns (including land and sea breezes) with fish availability, optimal fishing grounds, and safe navigation. This knowledge is often passed down through oral tradition.

- **Direct Observation:** Fishermen constantly observe the wind direction and strength. The shift from a land breeze to a sea breeze, or vice versa, signals a change in conditions that they interpret based on their experience.
- **Sailing:** For those still using sail-powered boats, the land and sea breezes are crucial for navigation to and from fishing grounds.
- **Local Ecological Knowledge:** Fishermen often connect the breezes to other environmental cues, such as water clarity, surface currents, and the presence of certain indicator species, to make decisions about where and when to fish.

In modern fisheries, while technology like GPS and weather forecasts are widely used, the understanding of local wind patterns like land and sea breezes can still be a valuable asset, especially for small-scale and artisanal fishermen who rely heavily on their knowledge of the marine environment.

Do you have any specific questions about how these breezes affect a particular aspect of marine fisheries?

Land and sea breezes

1. **What causes land and sea breezes?**
   a) Rotation of Earth
   b) Difference in salinity
   c) Differential heating of land and water
   d) Ocean currents
2. **During the day, which area heats up faster?**
   a) Sea
   b) Land
   c) Both equally
   d) Atmosphere
3. **Sea breeze blows from:**
   a) Land to sea
   b) Sea to land
   c) Mountain to valley
   d) Valley to mountain
4. **Land breeze usually occurs during:**
   a) Afternoon
   b) Morning
   c) Night
   d) Monsoon
5. **Sea breeze develops due to:**
   a) Land being hotter than sea
   b) Sea being hotter than land
   c) Rainfall over ocean
   d) Oceanic pressure
6. **Land breezes flow from:**
   a) Ocean to land
   b) Equator to poles
   c) Land to sea
   d) Upper air to surface

7. **Sea breeze helps to:**
   a) Increase land temperature
   b) Reduce land temperature
   c) Increase pressure
   d) Dry out soil
8. **Why is sea breeze stronger than land breeze in many regions?**
   a) Less friction over land
   b) Greater pressure gradient during the day
   c) Stronger Coriolis force at night
   d) Air is drier at night
9. **Land breeze is generally:**
   a) Strong and moist
   b) Weak and moist
   c) Strong and dry
   d) Weak and dry
10. **The air above land rises during the day because:**
   a) It becomes denser
   b) It cools down quickly
   c) It expands on heating
   d) It is replaced by cold air
11. **Sea breeze causes:**
   a) Cloud formation inland
   b) High pressure over land
   c) Warmer nights
   d) Suppression of convection
12. **Land and sea breezes are a type of:**
   a) Planetary wind
   b) Monsoon
   c) Local wind
   d) Trade wind
13. **Sea breeze can delay:**
   a) Sunset
   b) Evaporation
   c) Afternoon thunderstorms
   d) Onset of monsoon
14. **Sea breeze front is:**
   a) Boundary where breeze reverses
   b) Leading edge of cool marine air
   c) Jet stream over the ocean
   d) None of the above
15. **Clouds from sea breeze front can cause:**
   a) Heavy rainfall over sea
   b) Hailstorms
   c) Localized thunderstorms
   d) No precipitation
16. **Sea breeze helps aquatic ecosystems by:**
   a) Raising water temperature
   b) Enhancing surface mixing
   c) Blocking sunlight
   d) Reducing oxygen levels
17. **Land breeze impacts coastal waters by:**
   a) Warming surface water
   b) Initiating upwelling
   c) Stopping waves
   d) Causing cyclones

18. Upwelling due to land breeze results in:

a) Nutrient-poor water
b) Cooler surface waters and more nutrients
c) Warm surface water accumulation
d) Oxygen depletion

19. Fishermen prefer coastal upwelling zones because:

a) Clear skies
b) Less waves
c) Rich fish population
d) Shallow waters

20. Land breeze creates conditions favorable for:

a) Surface algae bloom
b) Pollutant accumulation
c) Deep-sea storms
d) Nutrient circulation

21. Typical time for sea breeze onset is:

a) Just after sunrise
b) Mid to late morning
c) Noon to afternoon
d) After sunset

22. Land breeze commonly reaches its peak intensity during:

a) Early morning
b) Late afternoon
c) Night around midnight
d) After rainfall

23. Land and sea breeze cycle repeats:

a) Weekly
b) Monthly
c) Every 24 hours
d) Seasonally

24. Duration of a typical land breeze is:

a) 12–18 hours
b) 4–6 hours
c) Entire day
d) 10 minutes

25. Sea breeze reaches farther inland when:

a) The sea is warmer
b) Land is hotter
c) Sky is cloudy
d) Night falls early

26. Land and sea breezes are most prominent in:

a) Equatorial regions
b) Polar regions
c) Coastal tropical and temperate zones
d) Deserts

27. Which city experiences strong sea breezes due to coastal proximity?

a) Delhi
b) Mumbai
c) Bhopal
d) Jaipur

28. Large inland lakes may also exhibit:

a) Global winds
b) Monsoon winds
c) Local land and lake breezes
d) Cyclonic circulations

29. Intensity of sea breeze is affected by:

a) Earth's orbit
b) Tidal cycle
c) Temperature contrast between land and water
d) Soil moisture

30. Deserts near coasts may still experience sea breeze due to:

a) High pressure zones
b) Ocean's thermal inertia
c) Rainfall
d) Forest cover

31. Thermal inertia refers to:

a) Rate of evaporation
b) Resistance to temperature change
c) Air pressure balance
d) Ocean salinity

32. Which property causes the ocean to heat more slowly than land?

a) Higher albedo
b) Greater thermal conductivity
c) Higher specific heat
d) Lower density

33. The direction of wind in a sea breeze is:

a) Horizontal from sea to land
b) Vertical from sea surface up
c) Downward from upper atmosphere
d) Circular motion

34. Sea breeze strength is typically weakest when:

a) Sky is clear
b) Land and sea temperatures are similar
c) Sun is overhead
d) Humidity is high

35. Land breeze is enhanced during:

a) Summer afternoons
b) Cloudy mornings
c) Clear, calm nights
d) Rainy nights

36. Sea breeze may benefit nearshore fish by:

a) Driving them to deeper water
b) Increasing water temperature
c) Circulating plankton-rich water
d) Blocking photosynthesis

37. Increased wind-induced mixing from sea breeze improves:

a) Water clarity
b) Thermal layering
c) Oxygen distribution
d) Salinity levels

38. Land breeze promotes:

a) Surface water accumulation
b) Warm current development
c) Coastal upwelling
d) Algal bloom suppression

39. Nutrient upwelling benefits:

a) Coral bleaching
b) Trophic food chains
c) Surface temperature rise
d) Oil layer stabilization

40. Fish populations are often denser in areas of:

a) Warm surface waters
b) Calm, nutrient-poor waters
c) Upwelling and nutrient enrichment
d) River mouths only

41. Which instrument is used to observe sea breezes?

a) Rain gauge
b) Barometer
c) Anemometer
d) Thermograph

42. On weather maps, sea breeze fronts can be identified by:

a) Sudden pressure rise
b) Drop in visibility
c) Wind direction shift and temperature drop
d) Increased humidity

43. Land breeze and sea breeze circulations are examples of:

a) Large-scale advection
b) Geostrophic flow
c) Convection cells
d) Cyclogenesis

44. Sea breeze strength is reduced on:

a) Clear, hot days
b) Windy days with prevailing offshore winds
c) Calm mornings
d) Humid afternoons

45. Sea breeze influences air quality by:

a) Trapping pollutants inland
b) Reducing coastal humidity
c) Bringing fresh marine air
d) Causing smog

46. Land-sea breeze circulation typically extends up to:

a) 100 meters
b) 1–2 km
c) 10 km
d) 20 km

47. The sea breeze penetrates farther inland when:

a) Water is colder
b) Temperature difference is higher
c) Mountains are nearby
d) Humidity is lower

48. Which type of weather is commonly associated with sea breeze convergence zones?

a) Snowfall
b) Local thunderstorms
c) Dust storms
d) Cyclones

49. Land-sea breeze circulation impacts are most evident during:

a) Equinoxes
b) Winter solstice
c) Calm summer days
d) Cloudy monsoon periods

50. Land and sea breezes are most useful in understanding:

a) Earthquake patterns
b) Long-term climate trends
c) Local weather variation and coastal ecology
d) Jet stream dynamics

Answer Key (s)

1	c	2	b	3	b	4	c	5	a	6	c	7	b	8	b	9	b	10	d
11	a	12	c	13	c	14	b	15	c	16	b	17	b	18	b	19	c	20	d
21	c	22	a	23	c	24	b	25	b	26	c	27	b	28	c	29	c	30	b
31	b	32	c	33	a	34	b	35	c	36	c	37	c	38	c	39	b	40	c
41	c	42	c	43	c	44	b	45	c	46	b	47	b	48	b	49	c	50	c

10

Cyclones and Anticyclones

These **cyclones** and **anticyclones**, the large-scale rotating weather systems, characterized by distinct patterns of atmospheric pressure and wind, exert a profound influence on ocean conditions that are directly relevant to fisheries. From the dramatic impacts of cyclones – with their powerful winds, heavy rainfall, and storm surges – to the more subtle but equally important effects of anticyclones on sea state and overall weather stability, understanding these systems is crucial for anyone involved in marine fisheries.

For those of you aiming to work in aquaculture, fishing operations, coastal management, or marine conservation, a solid grasp of cyclones and anticyclones will equip you to anticipate and respond to significant environmental changes. This knowledge will be invaluable for ensuring the safety of fishing fleets, managing aquaculture facilities in coastal areas, and understanding the broader ecological impacts on fish habitats and populations.

In this chapter, we will explore

- **The fundamental characteristics of cyclones and anticyclones:** We will unravel their pressure systems, wind patterns, and overall structure.
- **Their formation and life cycle:** We'll examine the atmospheric conditions that give rise to these powerful systems and how they evolve over time.
- **The direct and indirect impacts on the marine environment and fisheries:** This will include the effects of strong winds on sea state, the influence of rainfall and runoff on coastal salinity, and the devastating potential of storm surges on coastal ecosystems and infrastructure.1 We will also touch upon how anticyclones can lead to periods of calm weather that can also have implications for fishing activities.

Prepare to navigate the fascinating dynamics of these atmospheric giants and understand their significant role in shaping the marine world that underpins our fisheries. Let's begin our exploration of the whirlwinds of change!

10.1 Cyclones

A **cyclone**, at its core, is a vast, swirling vortex in the atmosphere, characterized by a central zone of significantly lower atmospheric pressure compared to its surroundings. This pressure difference creates a powerful gradient, driving air inwards towards the low-pressure center. The scale of these systems can be immense, often spanning hundreds to over a thousand kilometers in diameter. Unlike localized thunderstorms, cyclones are organized systems with sustained, rotating winds (Fig.10.1). Their intensity is a function of the pressure deficit at the center and the resulting pressure gradient, directly influencing wind speeds and the severity of associated weather phenomena. These phenomena include torrential rainfall, widespread cloud cover, high waves, and, in the case of intense tropical cyclones, devastating storm surges that can inundate coastal regions. The classification of a cyclone (e.g., tropical depression, tropical storm, hurricane, typhoon) depends on its sustained wind speed.

10.1.1 How Cyclones are Formed?

The genesis of a tropical cyclone is a complex process requiring a confluence of specific atmospheric and oceanic conditions.

- **Warm Ocean Waters:** The ocean acts as the cyclone's primary energy source. Sea surface temperatures of at least 26.5^{0}C (80^{0}F) need to extend to a considerable depth (around 50 meters). This warm layer provides the substantial heat and moisture necessary for the initial development and subsequent intensification. Evaporation from this warm surface adds significant water vapor to the lower atmosphere.
- **Atmospheric Instability:** The atmosphere must be inherently unstable, meaning that if a parcel of air is lifted, it will continue to rise because it remains warmer than its surroundings. This instability allows for the development of deep convective clouds (thunderstorms), which are the building blocks of a cyclone.
- **High Humidity in the Lower to Mid-Troposphere:** A moist mid-level atmosphere is crucial. If the air is too dry at these levels, the rising air parcels will evaporate before they can condense and form clouds, hindering the development of the storm's core.
- **Low Vertical Wind Shear:** Vertical wind shear, the change in wind speed or direction with height, must be minimal (typically less than 10 meters per second between the surface and the upper troposphere). Strong shear can tilt the developing storm's structure, disrupting the vertical transport of heat and moisture and potentially tearing it apart.

- **Pre-existing Low-Level Disturbance:** A pre-existing disturbance, such as a tropical wave (a westward-moving area of lower pressure that originates over Africa in the Atlantic), provides the initial spin and convergence needed to organize the developing thunderstorms.
- **Coriolis Force (Crucial Hemispheric Difference - Deep Dive):**
  - **Northern Hemisphere:** As air rushes towards the initial low-pressure area, the Earth's eastward rotation causes a deflection to the **right** due to the Coriolis Effect. This continuous rightward deflection of inflowing air results in a **counterclockwise** rotation around the developing low-pressure center. Without sufficient Coriolis force (very near the Equator), this sustained rotation cannot be established, which is why tropical cyclones rarely form within about 5 degrees of the Equator.
  - **Southern Hemisphere:** The Coriolis Effect in the Southern Hemisphere deflects moving air to the **left**. Consequently, the air drawn into the low-pressure center is continuously deflected to the left, resulting in a **clockwise** rotation of the developing cyclone. The same limitation regarding the proximity to the Equator applies here as well.

The process begins with the clustering of thunderstorms over warm waters. The rising air in these storms leads to condensation, releasing latent heat, which warms the surrounding air column. This warming aloft causes the air pressure at higher levels to rise, leading to divergence (outflow) of air at the top. This outflow helps to lower the surface pressure further, encouraging more warm, moist air to rise from the ocean surface, creating a positive feedback loop that can lead to the formation of a cyclone.

10.1.2 How Cyclones Get Intensified?

Once a tropical cyclone is born, its intensification hinges on maintaining and enhancing the energy transfer from the ocean and the organization of its structure:

- **Sustained Warm Water Contact:** As long as the cyclone remains over very warm waters, the ocean continues to provide a vast reservoir of heat and moisture, fueling the storm's engine. The deeper the layer of warm water, the more potential energy is available.
- **Efficient Moisture Inflow:** A well-defined low-level circulation allows for a continuous and large-scale inflow of warm, moist air towards the storm's center. This air rises within the thunderstorms, condenses, and releases latent heat.

- **Favorable Upper-Level Outflow (Enhanced):** At the top of the cyclone, divergence of air is crucial. This outflow, often enhanced by upper-level wind patterns, allows the surface pressure to drop further by removing mass from the column of air above the storm. A well-ventilated upper atmosphere is conducive to intensification.
- **Low Vertical Wind Shear:** Minimal shear allows the storm to remain vertically aligned, ensuring that the heat and moisture released at different altitudes contribute effectively to the overall intensification. High shear can disrupt this vertical coherence.

10.1.3 How Cyclones Dissipate?

The demise of a tropical cyclone occurs when its energy source is cut off or its organized structure is disrupted:

- **Movement Over Land:** Land surfaces lack the vast supply of warm moisture that oceans provide. As a cyclone moves over land, it is starved of its primary fuel. Additionally, friction with the land surface slows down the low-level winds, reducing the inflow of air towards the center and weakening the storm.
- **Movement Over Cooler Waters:** If the cyclone tracks over ocean waters that are significantly cooler than the required threshold, the rate of evaporation and heat transfer decreases dramatically, weakening the storm's intensity.
- **Encountering Strong Vertical Wind Shear:** High wind shear can tilt the cyclone's vertical axis, separating the low-level circulation from the upper-level outflow. This disruption inhibits the efficient upward transport of heat and moisture, leading to weakening.
- **Interaction with Mid-latitude Weather Systems:** When a tropical cyclone moves into higher latitudes, it can interact with fronts and other features of mid-latitude weather. This interaction can sometimes lead to the cyclone transitioning into an extratropical cyclone, a different type of storm system with different energy sources and structure, often accompanied by a broadening of the wind field and a change in the temperature structure.

10.1.4 Structure of Cyclones

A mature tropical cyclone exhibits a distinct and organized structure (Fig.10.1):

Fig. 10.1: Structure of cyclone

- **The Eye:** The eye is the relatively calm and clear central region of the cyclone. It is characterized by descending air, which leads to adiabatic warming and the suppression of clouds. The diameter of the eye can vary considerably, from just a few kilometers to over 100 kilometers. The pressure within the eye is the lowest in the entire storm.
- **The Eyewall:** Surrounding the eye is the eyewall, a ring of the most intense thunderstorms in the cyclone. Here, you find the strongest sustained winds, the heaviest rainfall, and the highest and most turbulent clouds. The eyewall is where the maximum energy conversion from the warm ocean occurs.
- **Rainbands:** Spiraling outwards from the eyewall are bands of thunderstorms known as rainbands. These bands can stretch for hundreds of kilometers and also contain heavy rain and strong, gusty winds. The intensity of the rain and wind generally decreases with distance from the center.
- **Outflow:** At the upper levels of the troposphere, air that has risen through the eyewall and rainbands flows outwards away from the storm's center. This upper-level divergence is crucial for maintaining the low pressure at the surface by removing air mass. The pattern of this outflow can sometimes indicate the future intensity changes of the cyclone.

10.1.5 Hemispheric Difference in Wind Circulation

- **Northern Hemisphere:** The winds in a cyclone spiral inwards in a **counterclockwise** direction.
- **Southern Hemisphere:** The winds in a cyclone spiral inwards in a **clockwise** direction.

This fundamental difference in the direction of rotation is a direct and significant consequence of the Coriolis effect acting oppositely in the two hemispheres. When looking at satellite imagery or weather maps, this rotational difference is a key identifier of the hemisphere in which the cyclone is located.

This more detailed explanation should provide a comprehensive understanding of cyclones.

10.2 Classification of Cyclones

Cyclones can be broadly classified based on their **origin** and **characteristics**. The two main categories are:

10.2.1 Tropical Cyclones

- **Origin:** Form over warm tropical or subtropical waters.
- **Energy Source:** Driven by the heat released when moist air condenses. They have a warm core.
- **Structure:** Typically have a well-defined center (eye), an eyewall of intense thunderstorms, and spiral rainbands.
- **Nomenclature:** Called hurricanes in the Atlantic and eastern Pacific, typhoons in the western Pacific, and cyclones in the Indian Ocean and South Pacific.
- **Classification by Intensity:** Tropical cyclones are further classified based on their maximum sustained wind speeds. Different regions have slightly varying scales, but generally include:
  - **Tropical Depression:** Maximum sustained winds less than 39 mph (63 km/h).
  - **Tropical Storm:** Maximum sustained winds between 39-73 mph (63-118 km/h) - at this stage, they receive a name.
  - **Hurricane/Typhoon/Severe Cyclonic Storm:** Maximum sustained winds of 74 mph (119 km/h) or higher. These are often further categorized (e.g., using the Saffir-Simpson Hurricane Wind Scale for hurricanes, with categories 1 to 5 based on wind speed and potential damage).

10.2.2 Tropical Cyclones (also known as Mid-latitude or Temperate Cyclones)

- **Origin:** Form in the mid and high latitudes (outside the tropics) along frontal boundaries where there are significant horizontal temperature gradients between air masses (baroclinic zones).
- **Energy Source:** Derive their energy from these temperature contrasts. They have a cold core.
- **Structure:** Typically larger than tropical cyclones, often lack a distinct eye, and are associated with fronts (cold, warm, and occluded).

- **Weather:** Bring widespread rain or snow, and can have strong winds, but generally not as intense as major tropical cyclones. Examples include Nor'easters.

10.2.3 Other Types of Cyclones (Less Common or on Different Scales)

- **Polar Cyclones:** Smaller and shallower low-pressure systems that form over polar regions.
- **Mesocyclones:** Rotating updrafts within severe thunderstorms over land, often associated with the development of tornadoes.

For fisheries students, the primary focus is usually on **tropical cyclones** due to their significant impact on coastal marine environments and the safety of fishing operations in tropical and subtropical regions. However, understanding **extratropical cyclones** is also relevant as they can affect weather and sea conditions in higher latitudes where fishing activities occur.

10.3 Intensity of Cyclones

Tropical cyclones are classified based on their maximum sustained wind speeds. Different regions around the world use slightly different scales and terminology, but the underlying principle is the same: higher wind speeds correspond to a more intense storm with the potential for greater damage. Here are some commonly used intensity scales:

10.3.1 Saffir-Simpson Hurricane Wind Scale (SSHWS) - Used in the North Atlantic and the northern Pacific east of the International Date Line

This scale classifies hurricanes (tropical cyclones with sustained winds of 74 mph / 119 km/h or higher) into five categories based on their maximum sustained wind speeds. It also provides a general idea of the potential property damage.

Category	Sustained Winds (mph)	Sustained Winds (km/h)	Potential Damage
1	74-95	119-153	Minimal: Damage primarily to unanchored mobile homes, poorly constructed signs, and vegetation. Some coastal flooding.
2	96-110	154-177	Moderate: Some roof, door, and window damage. Considerable damage to vegetation, mobile homes, and piers. Coastal flooding with evacuation of low-lying areas.
3	111-129	178-208	Extensive: Some structural damage to small residences and utility buildings. Mobile homes destroyed. Flooding near the coast can destroy smaller structures. (Major Hurricane)

Category	Sustained Winds (mph)	Sustained Winds (km/h)	Potential Damage
4	130-156	209-251	Catastrophic: More extensive structural damage with some complete roof failures. Extensive damage to doors and windows. Major damage to lower floors of coastal structures. (Major Hurricane)
5	≥ 157	≥ 252	Catastrophic: Complete roof failure and some complete building collapses. Major damage to all buildings near the coast. Massive evacuation may be required. (Major Hurricane)

10.3.2 Scales Used in Other Basins

Other regions use different terminology and slightly different wind speed thresholds:

- **Western North Pacific:** Uses terms like Tropical Storm, Typhoon, and Super Typhoon (for storms with very high sustained winds).
- **North Indian Ocean:** Uses terms like Cyclonic Storm, Severe Cyclonic Storm, Very Severe Cyclonic Storm, Extremely Severe Cyclonic Storm, and Super Cyclonic Storm.
- **Southwest Indian Ocean and Australian Region:** Use categories 1 to 5, similar in concept to the Saffir-Simpson scale but with different wind speed ranges.

Key takeaway for fisheries: The intensity of a tropical cyclone directly correlates with the potential for hazardous conditions at sea (high waves, strong winds) and along the coast (storm surge, flooding), all of which can significantly impact fishing operations, aquaculture facilities, and coastal ecosystems. Understanding these intensity scales helps in assessing the level of threat and the necessary precautions.

10.4 After Effects of Cyclones on Marine Fisheries

Cyclones can have significant and lasting after-effects on marine fisheries:

- **Habitat Destruction:** Storm surges and strong waves can damage or destroy coastal habitats like mangroves, coral reefs, and seagrass beds, which are vital for many fish species.
- **Changes in Water Quality:** Heavy rainfall and runoff can reduce salinity, increase turbidity, and introduce pollutants into coastal waters, stressing or killing marine life.

- **Dispersal of Fish and Larvae:** Strong currents can displace fish populations, disrupt spawning aggregations, and scatter fish larvae, affecting recruitment.
- **Damage to Fishing Infrastructure:** Fishing boats, gear, and aquaculture facilities can be damaged or destroyed, disrupting fishing activities and livelihoods.
- **Long-Term Impacts on Fish Populations:** The combined effects can lead to long-term declines in fish stocks, impacting fisheries productivity.
- **Increased Disease Outbreaks:** Stressed fish populations are more susceptible to disease, which can be exacerbated by poor water quality after a cyclone.
- **Changes in food web:** Changes in habitat, water quality, and plankton distribution can affect the entire food web.
- **Economic Impacts:** Loss of income for fishermen, damage to aquaculture, and reduced fish catches can have substantial economic consequences for coastal communities.
- **Safety at sea:** Cyclones create dangerous conditions at sea, preventing fishing activities.
- **Changes in ocean currents:** Cyclones can alter ocean currents, which can affect the distribution of nutrients and plankton, and thus the distribution of fish.
- **Mixing of the water column:** Cyclones can cause mixing of the water column, which can bring nutrients to the surface, but can also disrupt the stratification of the water column, which can affect the distribution of fish.
- **Changes in water temperature:** Cyclones can cause changes in water temperature, which can affect the distribution of fish.

10.5 Depressions

In meteorology, a **depression** (also known as a low-pressure system or simply a low) is an area where the atmospheric pressure is lower than that of the surrounding locations. Air tends to flow inwards towards these areas of lower pressure.Depending on their intensity and location, depressions can range from weak, localized lows that bring cloudy skies to intense systems like mid-latitude cyclones or tropical depressions (the initial stage of a tropical cyclone).

For our context, we'll primarily focus on the mid-latitude depressions that are common weather features outside of the tropics.

10.5.1 How They Look on Isobaric Charts

On an isobaric chart, a depression is represented by a series of roughly concentric, closed isobars with the lowest pressure indicated at the center.

- **Northern Hemisphere:** The isobars will encircle the low-pressure center. Due to the Coriolis effect, the winds around a mid-latitude depression in the Northern Hemisphere circulate **counterclockwise** and inwards.
- **Southern Hemisphere:** Similarly, the isobars will encircle the low-pressure center. However, in the Southern Hemisphere, the winds around a mid-latitude depression circulate **clockwise** and inwards due to the Coriolis effect acting in the opposite direction.

The closer the isobars are packed together, the steeper the pressure gradient, and consequently, the stronger the winds associated with the depression.

10.5.2 How Depressions form

Mid-latitude depressions typically form along **fronts**, which are boundaries between air masses with different temperatures and densities. The most common mechanism is **cyclogenesis** along a polar front:

- **Stationary Front:** Initially, there is often a stationary front separating cold polar air to the north and warmer subtropical air to the south.
- **Development of a Wave:** Under certain upper-level atmospheric conditions (like the presence of a jet stream and upper-level divergence), a "kink" or wave can develop along this front. This creates a localized area of lower pressure.
- **Warm and Cold Sectors:** As the wave amplifies, a distinct warm sector (the warmer air moving poleward) and a cold sector (the colder air pushing equatorward) develop. A warm front forms at the leading edge of the warm air, and a cold front forms at the leading edge of the cold air.
- **Cyclonic Circulation:** The pressure continues to fall at the center of the wave, and the Coriolis effect causes the air flowing towards this low center to rotate (counterclockwise in the Northern Hemisphere, clockwise in the Southern Hemisphere). This establishes a cyclonic circulation.

10.5.3 How Depressions Dissipate

Mid-latitude depressions typically dissipate through a process called **occlusion**:

- **Faster Movement of the Cold Front:** Cold fronts generally move faster than warm fronts.
- **Occluded Front Formation:** The faster-moving cold front eventually catches up to and overtakes the slower-moving warm front. The boundary created when this happens is called an **occluded front**.
- **Cut-off of Warm Air:** As the occlusion process continues, the warm sector is lifted aloft, and the cyclone becomes increasingly dominated by colder air at the surface.
- **Weakening Pressure Gradient:** Without the temperature contrast that initially fueled its development, the pressure gradient weakens, and the cyclonic circulation gradually decays. The depression "fills in" as the central pressure rises.
- **Friction:** Surface friction also plays a role in slowing down the winds and contributing to the dissipation over time, although it's not the primary cause of dissipation of the low-pressure system itself.

In summary, mid-latitude depressions are low-pressure systems that form along fronts, exhibit characteristic rotational patterns on isobaric charts (counterclockwise in the Northern Hemisphere, clockwise in the Southern Hemisphere), and typically dissipate through the process.

10.6 Differences between Cyclones and Depressions

Cyclones and depressions are both types of low-pressure systems that form over the ocean or land, but they differ in terms of their intensity, wind speed, and impact. Here's a brief overview:

10.6.1 Cyclones

- Also known as hurricanes or typhoons in different parts of the world
- Form over warm ocean waters when atmospheric conditions are favorable
- Characterized by strong winds, heavy rainfall, and rough seas
- Can cause significant damage, flooding, and loss of life

10.6.2 Depressions

- A low-pressure system with light to moderate winds and rainfall
- Can form over land or ocean

- Generally less destructive than cyclones, but can still cause some damage and disruptions

Here's a table summarizing the main differences:

Characteristics	Cyclone	Depression
Wind Speed	39-54 mph (63-87 km/h) or higher	Up to 38 mph (61 km/h)
Pressure	Very low atmospheric pressure	Low atmospheric pressure
Weather Conditions	Strong winds, heavy rainfall, thunderstorms	Light to moderate rainfall, cloudy skies
Impact	Significant damage, flooding, loss of life	Some damage, disruptions, but generally less severe
Formation	Forms over warm ocean waters	Forms over land or ocean

The key differences between cyclones and depressions lie in their wind speed, pressure, and impact. Cyclones are much more intense and destructive, while depressions are generally less severe.

10.7 Storms

A storm is a severe weather condition characterized by strong winds, heavy rainfall, thunder, lightning, and rough seas. Storms can form over land or ocean and can vary in intensity, duration, and impact. Storms can be classified into different types, such as:

- Thunderstorms
- Tropical cyclones (hurricanes, typhoons, cyclones)
- Winter storms
- Blizzards
- Dust storms

10.7.1 Characteristics of Storms

Some common characteristics of storms include:

- **Strong winds**: Storms are often accompanied by strong, gusty winds that can cause damage to structures, trees, and power lines.
- **Heavy rainfall**: Storms can bring heavy rainfall, leading to flash flooding, landslides, and other hazards.
- **Thunder and lightning**: Many storms are accompanied by thunder and lightning, which can be deadly.
- **Rough seas**: Storms over the ocean can cause rough seas, leading to coastal erosion, flooding, and damage to marine vessels.

- **Low atmospheric pressure**: Storms are often associated with low atmospheric pressure, which can cause a range of effects on the environment and human health.

10.7.2 Consequences of Storms on Fisheries

Storms can have significant impacts on fisheries, including:

- **Damage to fishing vessels and gear**: Strong winds and rough seas can damage or destroy fishing vessels, gear, and infrastructure.
- **Disruption to fishing activities**: Storms can force fishermen to suspend their activities, leading to economic losses and food security concerns.
- **Loss of livelihoods**: Storms can impact the livelihoods of fishermen and their families, particularly in coastal communities that depend heavily on fishing.
- **Impact on fish stocks**: Storms can disrupt fish habitats, leading to changes in fish populations and distributions.
- **Coastal erosion and habitat destruction**: Storms can cause coastal erosion and habitat destruction, affecting the nursery grounds and habitats of fish and other marine species.
- **Increased risk of accidents**: Storms can increase the risk of accidents and fatalities among fishermen, particularly if they are forced to venture out into rough seas.
- **Economic impacts**: Storms can have significant economic impacts on the fishing industry, including losses to infrastructure, vessels, and gear.

Overall, storms can have far-reaching consequences for fisheries, affecting not only the fishing industry but also the livelihoods and food security of coastal communities.

10.8 Anticyclones

An anticyclone is a high-pressure system where the atmospheric pressure is higher than the surrounding areas. Anticyclones are often associated with fair weather, clear skies, and light winds (Fig.10.2).

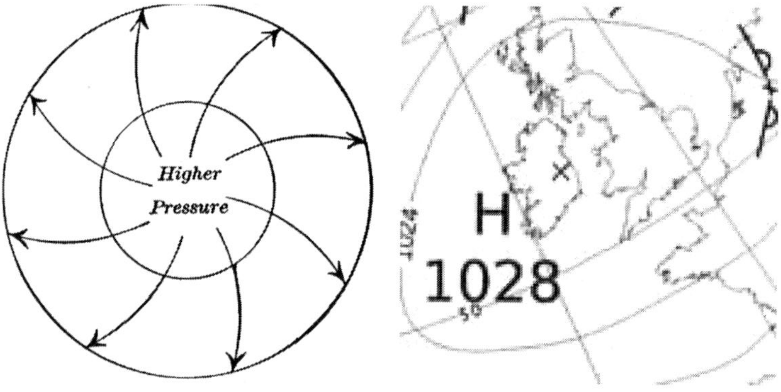

Fig. 10.2: Anticyclone

10.8.1 Characteristics of anticyclones

Some common characteristics of anticyclones include:

- **High atmospheric pressure**: Anticyclones are characterized by high atmospheric pressure at the center.
- **Fair weather**: Anticyclones are often associated with fair weather, clear skies, and light winds.
- **Stable air**: Anticyclones are characterized by stable air, which suppresses the formation of clouds and precipitation.
- **Light winds**: Anticyclones are often accompanied by light winds, which can lead to temperature inversions and fog.

10.8.2 Differences between cyclones and anticyclones

Here's a table summarizing the main differences between cyclones and anticyclones:

Characteristics	**Cyclone**	**Anticyclone**
Pressure	Low atmospheric pressure	High atmospheric pressure
Weather	Unstable, rainy, stormy	Fair, clear, stable
Winds	Strong, gusty winds	Light, gentle winds
Air movement	Air rises, creating clouds and precipitation	Air sinks, suppressing clouds and precipitation
Impact	Can cause damage, flooding, and loss of life	Generally associated with pleasant weather, but can lead to droughts and heatwaves
Formation	Forms over warm ocean waters or areas with strong atmospheric instability	Forms over areas with high pressure and stable atmospheric conditions

The key differences between cyclones and anticyclones lie in their pressure, weather, winds, and impact. Cyclones are associated with low pressure, unstable weather, and strong winds, while anticyclones are characterized by high pressure, fair weather, and light winds.

10.8.3 Significance of Anticyclones in Fisheries

Anticyclones can have both positive and negative impacts on fisheries, depending on the specific context and location. Here are some possible significance of anticyclones on fisheries:

- **Positive impacts**
  - **Calm seas**: Anticyclones can bring calm seas and favorable weather conditions, making it easier for fishermen to venture out and catch fish.
  - **Increased fish activity**: Stable weather conditions can lead to increased fish activity, making them more active and easier to catch.
  - **Improved fishing conditions**: Anticyclones can provide optimal fishing conditions, such as clear waters and good visibility, which can improve fishing success.
- **Negative impacts**
  - **Drought and reduced water levels**: Anticyclones can lead to droughts and reduced water levels in rivers, lakes, and reservoirs, affecting fish habitats and populations.
  - **Increased water temperature**: Anticyclones can cause increased water temperatures, which can stress fish populations and lead to changes in their distribution and behavior.
  - **Reduced nutrient upwelling**: Anticyclones can reduce nutrient upwelling, which can impact phytoplankton growth and ultimately affect the entire food chain.
  - **Impact on fish migration**: Anticyclones can disrupt fish migration patterns, which can affect the distribution and abundance of fish populations.
- **Other impacts**
  - **Changes in ocean currents**: Anticyclones can influence ocean currents, which can impact the distribution and migration of fish populations.
  - **Impact on coastal ecosystems**: Anticyclones can lead to changes in coastal ecosystems, such as increased growth of algae or changes in seagrass beds, which can affect fish habitats and populations.

Overall, the significance of anticyclones on fisheries depends on the specific context and location. While anticyclones can bring favorable fishing conditions, they can also have negative impacts on fish populations and ecosystems.

10.9 Sea Swell

Sea swell refers to a series of ocean waves that are generated by winds far out at sea. These waves can travel long distances without losing much energy and can cause significant wave activity when they reach the coast. Sea swells are often associated with storms or strong winds in the open ocean.

10.9.1 Characteristics of Sea Swell

- **Wave period**: Sea swells have a longer wave period than typical ocean waves, often ranging from 10 to 20 seconds or more.
- **Wave height**: Sea swells can cause significant wave heights, often exceeding 5-10 meters (16-33 feet) or more.
- **Wave direction**: Sea swells can approach the coast from various directions, depending on the wind direction and storm location.

10.9.2 Impacts of Sea Swell

- **Coastal erosion**: Sea swells can cause significant coastal erosion, especially during high-energy events.
- **Flooding**: Sea swells can lead to coastal flooding, especially during high tides or storm surges.
- **Damage to coastal infrastructure**: Sea swells can damage coastal infrastructure, such as seawalls, jetties, and buildings.

10.10 Storm Surges

A storm surge is a rise in sea level due to a storm, typically caused by a combination of strong winds and low atmospheric pressure. Storm surges can cause catastrophic coastal flooding and damage.

10.10.1 Characteristics of Storm Surges

- **Water level rise**: Storm surges can cause a significant rise in water levels, often exceeding 1-5 meters (3-16 feet) or more.
- **Wind-driven**: Storm surges are driven by strong winds that push water towards the coast.
- **Low atmospheric pressure**: The low atmospheric pressure associated with a storm can also contribute to the rise in water levels.

10.10.2 Impacts of Storm Surges

- **Coastal flooding**: Storm surges can cause catastrophic coastal flooding, leading to loss of life and property damage.
- **Damage to infrastructure**: Storm surges can damage or destroy coastal infrastructure, such as buildings, roads, and bridges.
- **Saltwater intrusion**: Storm surges can lead to saltwater intrusion into freshwater sources, affecting drinking water supplies and agriculture.

10.10.3 Factors that influence Storm Surges

- **Storm intensity**: The intensity of the storm, including wind speed and atmospheric pressure, can impact the severity of the storm surge.
- **Coastal geometry**: The shape and orientation of the coastline can influence the impact of the storm surge.
- **Tidal stage**: The timing of the storm surge with respect to the tidal cycle can affect the severity of the flooding.

10.10.4 Mitigation and preparedness

- **Early warning systems**: Early warning systems can provide critical information to communities at risk, allowing for evacuations and other preparations.
- **Coastal protection**: Coastal protection measures, such as seawalls and dunes, can help mitigate the impacts of storm surges and sea swells.
- **Evacuation plans**: Evacuation plans and emergency response strategies can help minimize the loss of life and property.

Overall, sea swells and storm surges are significant coastal hazards that can have devastating impacts on coastal communities and ecosystems. Understanding these phenomena is crucial for mitigating and preparing for these events.

10.11 Cyclone Warning and Alerts

The India Meteorological Department (IMD) plays a crucial role in issuing cyclone warnings and advisories to fishermen and other stakeholders. Here's a detailed explanation of how IMD alerts and warns fishermen about cyclones:

10.11.1 Cyclone Warning System

IMD has a well-established cyclone warning system that provides timely and accurate information to fishermen, coastal communities, and other stakeholders. The system involves a series of bulletins and warnings that are issued at regular intervals, depending on the severity of the cyclone.

10.11.2 Stages of Cyclone Warnings

IMD issues cyclone warnings in four stages:

- **Pre-Cyclone Watch**: This is the initial stage, where IMD issues a bulletin indicating the potential for cyclogenesis (formation of a cyclone) in the next 72 hours.
- **Cyclone Alert**: When a cyclone is likely to form within the next 48 hours, IMD issues a cyclone alert, advising fishermen and coastal communities to take necessary precautions.
- **Cyclone Warning**: When a cyclone is expected to affect the coast within the next 24 hours, IMD issues a cyclone warning, advising fishermen and coastal communities to take immediate action.
- **Post-Landfall Bulletin**: After the cyclone makes landfall, IMD issues a post-landfall bulletin, providing information on the cyclone's movement, wind speed, and rainfall.

10.11.3 Warning Dissemination

IMD disseminates cyclone warnings and advisories through various channels, including:

- **Bulletins**: IMD issues bulletins in English, Hindi, and local languages, which are broadcast through All India Radio (AIR) and Doordarshan (DD).
- **SMS and Mobile Alerts**: IMD sends SMS and mobile alerts to registered fishermen and other stakeholders, providing timely updates on cyclone warnings.
- **Social Media**: IMD uses social media platforms like Twitter and Facebook to disseminate cyclone warnings and advisories.
- **Fishermen's Mobile App**: IMD has developed a mobile app for fishermen, which provides real-time information on cyclone warnings, sea conditions, and fishing advisories.
- **Coastal Warning Dissemination System**: IMD has established a coastal warning dissemination system, which involves local authorities, fishermen's organizations, and other stakeholders in disseminating cyclone warnings to coastal communities.

10.11.4 Warning Content

IMD's cyclone warnings and advisories include essential information, such as:

- **Cyclone location and movement**: IMD provides information on the cyclone's current location, movement, and expected trajectory.
- **Wind speed and gusts**: IMD provides information on the cyclone's wind speed and gusts, which helps fishermen assess the risk.
- **Rainfall and storm surge**: IMD provides information on expected rainfall and storm surge, which helps fishermen and coastal communities prepare for potential flooding.
- **Fishing advisories**: IMD advises fishermen on safe fishing areas, potential hazards, and necessary precautions.

10.11.5 Impact on Fishermen

IMD's cyclone warnings and advisories have a significant impact on fishermen, enabling them to:

- **Take necessary precautions**: Fishermen can take necessary precautions, such as returning to safe harbors, securing their vessels, and protecting their gear.
- **Avoid risk**: Fishermen can avoid venturing into the sea during cyclonic conditions, reducing the risk of accidents and loss of life.
- **Plan their fishing activities**: Fishermen can plan their fishing activities according to the weather forecast, optimizing their catch and reducing losses.

In summary, IMD's cyclone warning system provides critical information to fishermen and coastal communities, enabling them to take necessary precautions and minimize the impact of cyclones.

Cyclones and Anticyclones

1. **A cyclone is defined as a system with:**
   a) High pressure and descending air b) Low pressure and ascending air
   c) High temperature and calm winds d) Cold core and low winds
2. **In the Northern Hemisphere, cyclonic winds rotate:**
   a) Clockwise b) Counter-clockwise
   c) Vertically upward d) From west to east
3. **Anticyclones are associated with:**
   a) Low pressure and stormy weather b) Rising air and clouds
   c) High pressure and descending air d) Monsoon winds

4. **In the Southern Hemisphere, the wind direction around an anticyclone is:**
   a) Clockwise
   b) Counter-clockwise
   c) Random
   d) Upward
5. **Tropical cyclones usually form over:**
   a) Mountains
   b) Deserts
   c) Warm ocean waters
   d) Ice-covered regions
6. **Which one of the following is NOT a feature of a cyclone?**
   a) Eye
   b) Strong winds
   c) Spiral rainbands
   d) High-pressure core
7. **The central calm area of a cyclone is called the:**
   a) Vortex
   b) Eye
   c) Core
   d) Center
8. **Anticyclones are typically associated with:**
   a) Cloudy and stormy weather
   b) Rainfall
   c) Clear skies and dry weather
   d) Thunderstorms
9. **Cyclones are most likely to intensify over:**
   a) Cold land
   b) Cold water
   c) Warm oceans
   d) Polar ice caps
10. **Which of the following is a tropical cyclone?**
    a) Tornado
    b) Typhoon
    c) Snowstorm
    d) Drought
11. **The Indian name for a tropical cyclone in the Bay of Bengal is:**
    a) Hurricane
    b) Typhoon
    c) Cyclone
    d) Twister
12. **Hurricanes are tropical cyclones in the:**
    a) Indian Ocean
    b) South Pacific
    c) Atlantic Ocean
    d) Mediterranean Sea
13. **A mid-latitude cyclone forms due to:**
    a) Tropical convergence
    b) Cold and warm air masses meeting
    c) Ocean currents
    d) Earthquakes
14. **Which cyclone category has wind speeds greater than 157 mph?**
    a) Category 3
    b) Category 2
    c) Category 5
    d) Category 1

15. Cyclones in the Pacific Ocean are called:

a) Typhoons
b) Hurricanes
c) Blizzards
d) Anticyclones

16. Extratropical cyclones differ from tropical cyclones in that they:

a) Originate over warm water
b) Form along frontal zones
c) Are stronger
d) Occur near the equator

17. The Saffir-Simpson scale is used to measure:

a) Rainfall in monsoons
b) Earthquake intensity
c) Cyclone wind speed
d) Humidity

18. Which term describes a weakening tropical cyclone?

a) Depression
b) Supercell
c) Dissipation
d) Intensification

19. Cyclone categories are primarily based on:

a) Rainfall
b) Duration
c) Wind speed
d) Temperature

20. A cyclone's spiral bands contain:

a) Dust clouds
b) Cold fronts
c) Heavy rainfall and thunderstorms
d) Calm weather

21. The storm surge in a cyclone is caused by:

a) Earthquakes
b) Intense winds and low pressure
c) Heavy rainfall alone
d) Inland flooding

22. Cyclones can result in all the following EXCEPT:

a) Flooding
b) Landslides
c) Drought
d) Coastal erosion

23. Anticyclones can lead to:

a) Snowstorms
b) Fog and air pollution buildup
c) Heavy rainfall
d) Tornadoes

24. A positive impact of cyclones is:

a) Soil erosion
b) Infrastructure damage
c) Rainfall in drought-prone areas
d) River flooding

25. Which is more likely during an anticyclonic condition?

a) Torrential rains
b) Storm surge
c) Clear skies and temperature inversion
d) Wind shear

26. Cyclones affect fisheries by:

a) Enhancing productivity
b) Damaging marine habitats and disrupting catch
c) Preventing water mixing
d) Causing nutrient depletion

27. Which sector is least affected directly by cyclones?

a) Agriculture
b) Fisheries
c) Transport
d) Astronomy

28. Cyclones may damage coral reefs due to:

a) Sudden cooling
b) High pressure
c) Strong wave action
d) Low winds

29. Anticyclones can result in:

a) Persistent rainfall
b) Droughts during dry seasons
c) Strong thunderstorms
d) Lightning storms

30. High-pressure systems inhibit rainfall because they:

a) Pull in ocean air
b) Cause air to rise and condense
c) Push air downward, preventing cloud formation
d) Increase humidity

31. Which instrument is commonly used to detect and monitor cyclones?

a) Thermometer
b) Barometer
c) Radar
d) Hygrometer

32. Which government agency issues cyclone warnings in India?

a) ISRO
b) CSIR
c) IMD
d) ICAR

33. One major indicator of an approaching cyclone is:

a) A rapid increase in temperature
b) Sudden drop in atmospheric pressure
c) Wind blowing from sea to land
d) Cloudless sky

34. Which of the following causes the most fatalities during a cyclone?

a) Wind gusts
b) Thunder
c) Storm surge and flooding
d) Lightning

35. The Coriolis effect causes cyclones to rotate differently in each hemisphere due to:

a) Gravity
b) Earth's orbit
c) Earth's rotation
d) Sun's radiation

36. Which is NOT a condition necessary for cyclone formation?

a) Warm sea surface temperature
b) High humidity
c) Strong vertical wind shear
d) Low-level convergence

37. Tropical cyclones derive energy primarily from:

a) Solar radiation
b) Cold ocean currents
c) Latent heat from condensation
d) Earth's magnetic field

38. The movement of tropical cyclones is influenced by:

a) Ocean salinity
b) Jet streams and trade winds
c) Local vegetation
d) Mountain ranges

39. Which warning signal number indicates the highest level of danger from a cyclone in India?

a) 1
b) 3
c) 5
d) 10

40. For inland fisheries, cyclones primarily impact through:

a) Habitat expansion
b) Debris inflow and water contamination
c) Cooling of water bodies
d) Increased oxygenation

41. Which Indian state is most frequently hit by tropical cyclones?

a) Gujarat
b) Maharashtra
c) Odisha
d) Punjab

42. Cyclone Fani (2019) made landfall in which Indian state?

a) Andhra Pradesh
b) West Bengal
c) Odisha
d) Tamil Nadu

43. Which international body helps in cyclone monitoring and warning coordination in Asia?

a) IPCC
b) SAARC
c) WMO
d) FAO

44. Cyclones can result in fish kills mainly due to:

a) UV radiation
b) Nutrient exhaustion
c) Oxygen depletion and water turbulence
d) Algal blooms

45. Which marine structure is often damaged by cyclonic waves?

a) Silos
b) Ponds
c) Fish cages and pens
d) Water wheels

46. Cyclonic rainfall is usually:

a) Light and continuous
b) Sporadic and icy
c) Heavy and intense over a short period
d) Dry and windy

47. Which of these is commonly observed before a cyclone hits?

a) Increase in atmospheric pressure
b) Sudden warming of sea water
c) Drop in temperature and calm sea
d) Increased solar radiation

48. Cyclones in the Indian Ocean region are named by:

a) WMO Headquarters
b) Indian Government only
c) Member countries of the WMO Panel
d) UN Climate Group

49. Which is a direct impact of cyclones on aquaculture farms?

a) Genetic drift
b) Enhanced productivity
c) Destruction of enclosures and escape of stock
d) Salinity reduction

50. The calm region in the center of a cyclone is known for:

a) High wind speed
b) Light rain and lightning
c) Clear skies and still air
d) Extreme turbulence

Answer Key (s)

1	b	2	b	3	c	4	a	5	c	6	d	7	b	8	c	9	c	10	b
11	c	12	c	13	b	14	c	15	a	16	b	17	c	18	c	19	c	20	c
21	b	22	c	23	b	24	c	25	c	26	b	27	d	28	c	29	b	30	c
31	c	32	c	33	b	34	c	35	c	36	c	37	c	38	b	39	d	40	b
41	c	42	c	43	c	44	c	45	c	46	c	47	c	48	c	49	c	50	c

11

Weather Forecasting

From the farmer deciding when to sow seeds to the traveler planning their journey, from the pilot navigating the skies to the energy provider managing demand, the state of the atmosphere profoundly influences our daily lives and myriad human activities. The ability to anticipate future atmospheric conditions, a practice we know as **weather forecasting**, is therefore not merely an academic pursuit but a practical necessity (Fig. 11.1).

Fig. 11.1: Weather forecasting

This chapter will look into the fascinating world of weather forecasting, exploring its historical roots, the fundamental scientific principles that underpin it, and the sophisticated tools and techniques employed by modern meteorologists. We will journey through the process of observing the current state of the atmosphere, the computational power used to model its complex dynamics, and the various ways forecasts are disseminated and interpreted.

Understanding the science behind weather forecasting empowers us to appreciate its capabilities and limitations. It allows us to make more informed decisions in the face of an ever-changing atmospheric environment. Join us as we unravel the methods used to predict the weather, from simple observations to cutting-edge numerical models, and gain insight into the continuous advancements shaping the future of this vital scientific discipline.

11.1 Types of Weather Forecast

There are several ways to categorize weather forecasts, most commonly by the **time range** they cover. Here are the main types:

11.1.1 Nowcasting

- **Time Range:** Present to 0-6 hours ahead. Some definitions narrow this to within the next 2 hours.
- **Focus:** Detailed description of the current weather and very short-term predictions of its evolution.
- **Methods:** Relies heavily on real-time observations from radar, satellites, and surface stations, extrapolated forward in time. Often incorporates mesoscale numerical weather models.
- **Usefulness:** Critical for issuing immediate warnings for severe weather (e.g., thunderstorms, heavy rainfall), aviation, and short-term planning of outdoor activities.
- **Example:** Predicting that a current heavy rain shower will continue to move northeast over the next hour, affecting a specific locality.

11.1.2 Short-Range Forecast

- **Time Range:** 1 to 3 days (up to 72 hours).
- **Focus:** Predicting day-to-day weather conditions, including temperature, precipitation, wind, and cloud cover.
- **Methods:** Primarily based on numerical weather prediction (NWP) models, which use computer simulations of the atmosphere. Meteorologists analyze and interpret model outputs, incorporating their expertise and local knowledge.
- **Usefulness:** Essential for daily planning, agriculture, transportation, and many other sectors.
- **Example:** Forecasting sunny skies and a high of 30°C for tomorrow.

11.1.3 Medium-Range Forecast

- **Time Range:** 3 to 10 days. Some sources extend this up to 2 weeks.
- **Focus:** Providing a general overview of weather trends beyond the immediate short-term. Accuracy decreases with increasing time.
- **Methods:** Primarily relies on global NWP models. Ensemble forecasting (running multiple simulations with slightly different initial conditions) becomes more important to assess the range of possible outcomes and uncertainty.

- **Usefulness:** Helpful for medium-term planning in various industries, including agriculture, energy, and logistics.
- **Example:** Indicating a likelihood of a wet spell arriving in the region around the end of next week.

11.1.4 Extended-Range Forecast

- **Time Range:** Beyond 10 days and up to a month (some define it as between medium-range and long-range, roughly days 10 to 30).
- **Focus:** Identifying broad-scale weather patterns and anomalies (e.g., above or below average temperatures, wetter or drier than normal conditions).
- **Methods:** Relies on global climate models and statistical methods that consider large-scale patterns like sea surface temperatures and atmospheric oscillations.
- **Usefulness:** Useful for long-term planning and for anticipating potential deviations from seasonal norms.

11.1.5 Long-Range Forecast (Seasonal Outlook)

- **Time Range:** One month to several seasons (e.g., 3-6 months or even a year).
- **Focus:** Predicting overall trends and probabilities of certain conditions (e.g., a warmer than average winter, a wetter than average monsoon). Does not provide day-to-day specifics.
- **Methods:** Based on complex global climate models that simulate the interactions between the atmosphere, oceans, land, and ice. Statistical analysis of historical data and teleconnections (long-distance relationships between weather patterns) are also used.
- **Usefulness:** Crucial for strategic planning in agriculture, water resource management, and energy sectors.
- **Example:** Forecasting a higher probability of a warmer and drier than normal summer for a particular region.

Additionally, forecasts can be categorized by their **purpose** or the **phenomena** they predict, such as:

- **Severe Weather Forecasts/Warnings:** Focused on hazardous weather like hurricanes, tornadoes, blizzards, and heatwaves.
- **Aviation Forecasts:** Tailored for the needs of the aviation industry, including wind, visibility, and icing conditions.

- **Marine Forecasts:** Providing information relevant to marine activities, such as wave height, sea state, and winds.
- **Agricultural Forecasts (Agromet Advisories):** Specific to farming, including rainfall, temperature, humidity, and their impact on crops.

The time-range categorization is the most common way to differentiate types of weather forecasts.

11.2 Weather Forecasting for Agriculture

Weather plays a pivotal role in agriculture, influencing everything from planting and growth to harvesting and the prevalence of pests and diseases. Accurate and timely weather forecasts tailored to agriculture are therefore indispensable for farmers to make informed decisions, optimize productivity, and mitigate risks (Fig. 16.2). This specialized area of weather forecasting is often referred to as **agrometeorological forecasting** or the provision of **agromet advisories**. **Key features of agricultural weather forecasts are:**

Agricultural weather forecasts go beyond general public forecasts by often including parameters critical for farming operations:

- **Rainfall:** Amount, intensity, timing, and spatial distribution are crucial for irrigation scheduling, planting, harvesting, and assessing the risk of floods or droughts.
- **Temperature:** Minimum and maximum temperatures affect crop development stages, the likelihood of frost damage, and the activity of pests and diseases. Soil temperature is also important for germination.
- **Humidity:** High humidity can favor the spread of fungal diseases, while low humidity can increase evapotranspiration rates and the need for irrigation.
- **Wind:** Speed and direction are important for planning pesticide and fertilizer applications (to avoid drift), assessing evapotranspiration, and understanding the potential for lodging (falling over) in crops.
- **Solar Radiation:** Affects photosynthesis and crop growth rates.
- **Evapotranspiration (ET):** An estimate of water loss from the soil and plants, vital for irrigation management.
- **Dew Point:** Indicates the likelihood of dew formation, which can influence the onset and spread of certain plant diseases.
- **Extreme Weather Events:** Warnings for frost, heat waves, hailstorms, strong winds, and other severe weather that can damage crops or livestock.

Fig. 11.2: Weather forecasting for agriculture

11.2.1 Types of Agricultural Weather Forecasts (by Time Range)

Similar to general weather forecasts, agricultural forecasts are also categorized by their lead time:

- **Nowcasting (0-6 hours):** Crucial for immediate decisions like whether to proceed with spraying pesticides or to take shelter from an approaching thunderstorm. Relies heavily on real-time data.
- **Short-Range (1-3 days):** Helps in planning daily farm activities such as irrigation, fertilizer application, and scheduling labor.
- **Medium-Range (3-10 days):** Useful for more strategic planning, like deciding on the timing of sowing or harvesting, and managing water resources over the coming week.
- **Extended-Range (10 days to a month):** Provides insights into broader trends, assisting in decisions about crop management and resource allocation over a longer period.
- **Long-Range/Seasonal Outlook (1 month to several seasons):** Invaluable for making decisions about which crops to plant in a particular season, planning for potential water availability (especially important for monsoon-dependent regions like India), and anticipating potential pest or disease outbreaks based on expected seasonal conditions.

11.2.2 Methods and Tools for Agricultural Weather Forecasting

Generating these specialized forecasts involves a combination of:

- **Traditional Meteorological Observations:** Data from surface weather stations, radiosondes (weather balloons), buoys, and aircraft provide the foundational information.
- **Satellite Imagery:** Offers a broad view of cloud cover, vegetation health, and land surface temperature, which are all relevant to agriculture.

- **Radar:** Detects precipitation intensity and movement, crucial for short-range warnings of heavy rainfall or hail.
- **Numerical Weather Prediction (NWP) Models:** Sophisticated computer models that simulate the atmosphere based on physical laws. These models are increasingly incorporating land surface processes relevant to agriculture, such as soil moisture.
- **Agrometeorological Models:** These models go a step further by integrating weather data with information about specific crops (their growth stages, water requirements, susceptibility to pests, etc.) to provide tailored advisories.
- **Statistical Methods:** Analyzing historical weather data and crop yields to identify relationships and make probabilistic forecasts.
- **Artificial Intelligence (AI) and Machine Learning (ML):** Increasingly used to analyze large datasets, improve the accuracy of forecasts, and develop more localized predictions.

11.2.3 Agromet Advisory Services

In many countries, including India, weather forecasts are translated into actionable advice for farmers through **Agrometeorological Advisory Services (AAS)**. These services typically involve:

- **Preparing Agromet Bulletins:** These bulletins, often issued 2-3 times a week, contain weather forecasts relevant to agriculture for a specific region, along with crop-specific advisories.
- **Crop-Specific Recommendations:** Based on the forecast, advisories might suggest the optimal timing for sowing, irrigation, application of fertilizers and pesticides, harvesting, and other farm operations.
- **Pest and Disease Warnings:** Predicting conditions favorable for the outbreak of pests and diseases and advising on preventive or control measures.
- **Warnings for Extreme Weather:** Alerting farmers to impending severe weather events and suggesting measures to protect their crops and livestock.
- **Dissemination:** Advisories are often disseminated through various channels like mobile SMS, radio, television, newspapers, and increasingly through dedicated mobile apps and web portals in local languages. In India, the India Meteorological Department (IMD) and agricultural universities play a significant role in this.

11.2.4 Importance for Agriculture

Accurate agricultural weather forecasts and effective agromet advisory services can lead to:

- **Optimized Planting and Harvesting:** Aligning these activities with favorable weather conditions.
- **Efficient Water Management:** Scheduling irrigation based on predicted rainfall and evapotranspiration.
- **Timely Application of Inputs:** Applying fertilizers and pesticides when they will be most effective and least likely to be washed away.
- **Reduced Crop Losses:** Taking protective measures against adverse weather, pests, and diseases.
- **Improved Resource Use Efficiency:** Making better decisions regarding labor, machinery, and other resources.
- **Enhanced Productivity and Income:** Ultimately leading to higher yields and better economic returns for farmers.

In conclusion, weather forecasting for agriculture is a specialized and vital service that combines meteorological science with agricultural knowledge to support farmers in making weather-smart decisions throughout the crop cycle. As technology advances, these forecasts are becoming more accurate, localized, and tailored to the specific needs of the agricultural community.

11.3 Weather Warnings and Alerts for Agriculture

Weather warnings and alerts are crucial for farmers as they help protect crops, livestock, and farm infrastructure from potentially damaging weather events. These warnings enable farmers to take proactive measures to minimize losses and ensure their safety (Fig. 11.3). Here are some common types of weather warnings and alerts relevant to agriculture:

11.3.1 Rainfall Warnings

- **Heavy Rainfall Warning:** Alerts farmers to the possibility of significant rainfall that could lead to waterlogging in fields, soil erosion, and damage to crops, especially during flowering or harvesting stages. Advisories might include draining excess water, postponing irrigation, or harvesting mature crops quickly.
- **Excessive Rainfall/Flash Flood Warning:** Indicates a high risk of sudden and intense rainfall that can cause widespread flooding, damaging crops, washing away topsoil, and posing a threat to livestock and farm

buildings. Farmers might be advised to move livestock to higher ground, secure farm equipment, and avoid low-lying areas.

- **Deficient Rainfall/Drought Alert:** Warns of prolonged dry spells that could lead to water stress in crops, reduced yields, and scarcity of water for irrigation and livestock. Advisories could include implementing water-saving irrigation techniques, selecting drought-tolerant crops, or managing livestock water resources carefully.

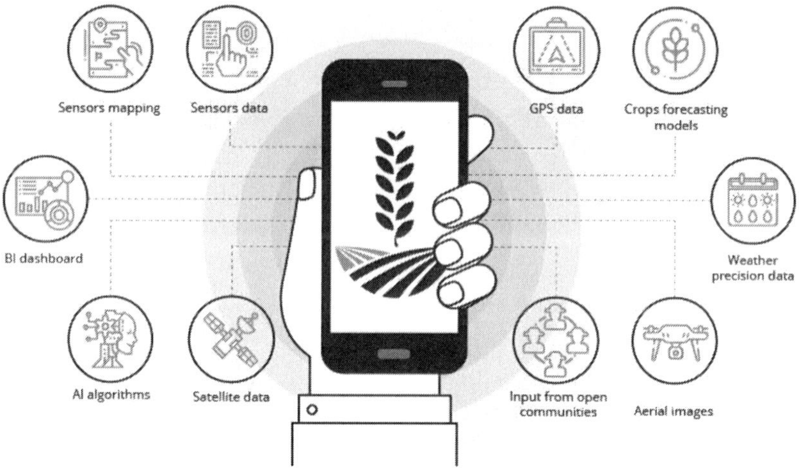

Fig. 11.3: Weather warnings and alerts for agriculture

11.3.2 Temperature Warnings

- **Frost Warning:** Alerts farmers to the risk of freezing temperatures that can damage sensitive crops, especially during early growth stages or flowering.3 Protective measures like covering plants, using smudge pots, or irrigating before frost might be recommended.
- **Heat Wave Warning:** Indicates dangerously high temperatures that can cause heat stress in crops and livestock, leading to reduced productivity and even mortality. Farmers might be advised to provide shade and ample water for livestock, adjust irrigation schedules for crops, and avoid working outdoors during peak heat hours.

11.3.3 Wind Warnings

- **Strong Wind Warning:** Alerts to the possibility of high winds that can cause lodging in crops (especially tall ones like maize or sugarcane), damage to young plants, and make spraying of pesticides or fertilizers ineffective. Farmers might be advised to provide support to vulnerable crops or postpone spraying.

- **Storm/Cyclone Warning:** Indicates the potential for very strong winds and heavy rainfall associated with storms or cyclones, which can cause widespread damage to crops, trees, and farm infrastructure. Farmers would typically be advised to secure buildings, move livestock to safety, and take shelter.

11.3.4 Hail Warnings

- **Hailstorm Warning:** Alerts to the risk of hailstorms that can cause significant physical damage to crops, fruits, and vegetables. Protective measures are limited but might include using netting in some high-value crop situations.

11.3.5 Other Warnings

- **Thunderstorm and Lightning Warnings:** Primarily for the safety of farmers and farmworkers, advising them to seek shelter to avoid lightning strikes.
- **Pest and Disease Outbreak Warnings:** While not strictly weather warnings, these are often linked to weather conditions favorable for pest and disease proliferation.4 Farmers are alerted to monitor their crops and take preventative measures.

11.4 Dissemination of Warnings and Alerts

Timely dissemination is crucial for these warnings to be effective. Common methods include:

- **Mobile SMS alerts:** Direct messages to farmers' mobile phones.
- **Radio and Television Broadcasts:** Reaching a wider audience, especially in rural areas.
- **Dedicated Mobile Apps:** Providing detailed weather information and advisories. Examples in India include 'Meghdoot' and 'Mausam'.
- **Web Portals:** Offering comprehensive weather information and agricultural advisories. The India Meteorological Department (IMD) provides these services through its website and the 'Mausamgram' portal.
- **Kisan Call Centres:** Providing information and answering queries in local languages.
- **Social Media Platforms:** Used by some Agromet Field Units (AMFUs) to disseminate information quickly.

- **Gram Panchayat Level Weather Forecasting (GPLWF):** Initiatives to provide forecasts at the village level through platforms like e-Gramswaraj and the Meri Panchayat app.9

In India, the **India Meteorological Department (IMD)** plays a central role in generating weather forecasts and warnings, which are then used by Agromet Field Units (AMFUs) to prepare and disseminate crop-specific advisories.10 These advisories often include warnings about impending adverse weather and suggest actions farmers should take.

11.5 Weather Forecasting Methods and Techniques

Weather forecasting involves using various methods and techniques to predict future weather conditions. These methods can be broadly categorized into traditional methods, modern methods, and advanced techniques.

11.5.1 Traditional Methods

- **Observational Method**: This method involves observing current weather conditions and using that information to predict future weather patterns. For example, if it is raining in a particular area, it is likely to continue raining in the same area for a short period. This method relies on the experience and expertise of forecasters to analyze current weather conditions and make predictions.
- **Persistence Method**: This method assumes that the current weather pattern will persist into the future. For example, if it is sunny today, it is likely to be sunny tomorrow. This method is simple and effective for short-term forecasting but can be less accurate for longer-term forecasts.
- **Climatological Method**: This method uses historical climate data to predict future weather patterns. For example, if a particular region typically experiences a certain type of weather during a particular time of year, it is likely to experience similar weather conditions in the future. This method is useful for long-term climate predictions but can be less accurate for short-term weather forecasts.

11.5.2 Modern Methods

- **Numerical Weather Prediction (NWP) Models**: These models use complex algorithms and large datasets to simulate the behavior of the atmosphere and predict future weather patterns. NWP models take into account various atmospheric and terrestrial factors, such as temperature, humidity, wind speed, and atmospheric pressure, to predict future weather conditions. These models are highly accurate and can provide detailed forecasts for specific locations.

- **Satellite Imagery**: Satellites in orbit around the Earth provide images of cloud patterns, precipitation, and other weather phenomena, which can be used to predict weather patterns. Satellite imagery can be used to track storms, monitor weather patterns, and provide early warnings for severe weather events.
- **Radar**: Radar systems use radio waves to detect precipitation and other weather phenomena, providing detailed information on current weather conditions. Radar data can be used to track storms, predict precipitation patterns, and provide early warnings for severe weather events.
- **Weather Stations**: Weather stations provide real-time data on temperature, humidity, wind speed, and other weather conditions, which can be used to predict future weather patterns. Weather stations can be located on the ground or on buoys in the ocean.

11.5.3 Advanced Techniques

- **Data Assimilation**: This technique involves combining model output with observational data to produce the best possible forecast. Data assimilation can help to improve the accuracy of NWP models by incorporating real-time data into the forecasting process.
- **Model Output Statistics (MOS)**: MOS involves using statistical techniques to combine the output of NWP models with observational data to improve forecast accuracy. MOS can help to reduce the bias in NWP models and provide more accurate forecasts.
- **Ensemble Forecasting**: This method involves generating multiple forecasts using different models or model versions, and then combining the results to produce a single forecast. Ensemble forecasting can help to quantify the uncertainty in forecasts and provide more accurate predictions.
- **Nowcasting**: Nowcasting involves using current weather conditions and model output to predict weather patterns over short time scales (typically 0-2 hours). Nowcasting is useful for predicting severe weather events, such as thunderstorms or tornadoes.

11.5.4 Benefits and Limitations

Each method and technique has its benefits and limitations. For example:

- NWP models are highly accurate but can be limited by the complexity of thevatmosphere and the quality of the input data.

- Satellite imagery provides valuable information on cloud patterns and precipitation but can be limited by the resolution and accuracy of the images.
- Radar systems provide detailed information on current weather conditions but can be limited by their range and resolution.

By combining traditional methods with modern technologies and advanced techniques, forecasters can provide accurate and reliable forecasts that support decision-making in a wide range of applications.

11.6 Synoptic Charts, Preparation and Use in Weather Forecasting

11.6.1 Synoptic Charts

A synoptic chart, also known as a weather map, is a graphical representation of weather conditions over a large area, typically at a specific time. These charts are used to analyze and forecast weather patterns by displaying various weather elements such as:

- **Isobars**: lines of equal atmospheric pressure
- **Fronts**: boundaries between different air masses
- **Wind direction and speed**: indicated by arrows or barbs
- **Temperature and dew point**: shown as numbers or isotherms
- **Precipitation**: indicated by symbols or shading

11.6.2 Use of Synoptic Charts

Synoptic charts are essential tools for weather forecasting, allowing forecasters to:

- **Analyze current weather patterns**: Synoptic charts provide a snapshot of current weather conditions, enabling forecasters to identify patterns and trends.
- **Identify weather systems**: Synoptic charts help forecasters identify and track weather systems such as high and low-pressure systems, fronts, and tropical cyclones.
- **Predict future weather patterns**: By analyzing the movement and development of weather systems, forecasters can predict future weather patterns and make informed decisions.
- **Issue weather warnings and advisories**: Synoptic charts are used to identify areas that may be affected by severe weather, enabling forecasters to issue timely warnings and advisories.

11.6.3 Limitations of Synoptic Charts

While synoptic charts are a valuable tool for weather forecasting, they have some limitations:

- **Data quality and availability**: The accuracy of synoptic charts depends on the quality and availability of weather data, which can be limited in some areas.
- **Complexity of weather patterns**: Synoptic charts can be difficult to interpret, especially when dealing with complex weather patterns or multiple weather systems.
- **Limited spatial and temporal resolution**: Synoptic charts typically provide a broad overview of weather patterns, but may not capture small-scale or rapidly changing weather phenomena.
- **Model dependence**: Synoptic charts often rely on model output, which can bc subject to errors and biases.

11.6.4 Modern Developments

Despite these limitations, synoptic charts remain an essential tool for weather forecasting. Modern developments, such as:

- **High-resolution models**: provide more detailed and accurate forecasts
- **Advanced data assimilation techniques**: improve the accuracy of synoptic charts
- **Automated chart production**: enable rapid production and dissemination of synoptic charts. Have enhanced the utility of synoptic charts in weather forecasting.

By understanding the strengths and limitations of synoptic charts, forecasters can effectively use these tools to analyze and predict weather patterns, ultimately providing better weather forecasts and warnings.

11.7 Weather Forecasting based on Satellite Pictures

Weather forecasting using satellite imagery is a crucial tool for predicting weather patterns, especially in rapidly changing conditions. Satellites in orbit around the Earth provide images of cloud patterns, precipitation, and other weather phenomena, which can be used to track storms, monitor weather patterns, and issue timely warnings.

11.7.1 Types of Satellites Used

- **Geostationary Satellites**: These satellites remain stationary relative to a specific point on Earth, providing continuous coverage of weather patterns. Examples include NOAA GOES, EUMETSAT Meteosat, JMA Himawari, and KAI Geo-KOMPSAT.
- **Polar-Orbiting Satellites**: These satellites orbit the Earth in a north-south direction, providing high-resolution images of the planet's surface and atmosphere. Examples include NASA's Aqua and Terra satellites.

11.7.2 How Satellite Imagery is Used

- **Tracking Weather Systems**: Satellite imagery helps forecasters track weather systems such as high and low-pressure systems, fronts, and tropical cyclones.
- **Monitoring Precipitation**: Satellites can detect precipitation an.d9 provide real-time data on rain and snow.
- **Predicting Weather Patterns**: By analyzing satellite imag.ery, forecasters can predict future weather patterns and issue timely warnjings.

11.7.3 Techniques Used

- **Artificial Neural Networks (ANNs)**: ANNs can be trained to analyze satellite imagery and predict weather patterns, such as rainfall and precipitation.
- **Data Assimilation**: This technique combines model output with observational data, including satellite imagery, to produce accurate forecasts.
- **Nowcasting**: This technique uses current weather conditions and model output to predict weather patterns over short time scales.

11.7.4 Benefits and Applications

- **Improved Forecast Accuracy**: Satellite imagery provides valuable data for predicting weather patterns, especially in rapidly changing conditions.
- **Timely Warnings**: Satellite imagery helps forecasters issue timely warnings for severe weather events, such as hurricanes and storms.
- **Global Coverage**: Satellites provide global coverage, enabling forecasters to track weather patterns and issue warnings for areas without radar systems.

11.7.5 Limitations and Challenges

- **Data Quality and Availability**: The accuracy of satellite imagery depends on the quality and availability of data, which can be limited in some areas.
- **Complexity of Weather Patterns**: Satellite imagery can be difficult to interpret, especially when dealing with complex weather patterns or multiple weather systems.
- **Model Dependence**: Satellite imagery often relies on model output, which can be subject to errors and biases.

11.8 New Initiatives for Operational Meteorological Advisories in the Fisheries Sector

The fisheries sector is highly susceptible to weather and climate variability, which can significantly impact fish production, livelihoods, and food security (Fig. 11.4). To enhance the resilience of the fisheries sector, new initiatives are necessary to improve operational meteorological advisories. Some potential initiatives include:

Fig. 11.4: Operational meteorological advisories in fisheries

- **Customized Weather Forecasting Services**: Developing tailored weather forecasting services that cater to the specific needs of the fisheries sector, including predictions of sea state, wind, waves, and ocean currents.
- **Marine Weather Warning Systems**: Establishing marine weather warning systems that provide timely and accurate warnings to fishermen and other stakeholders about severe weather events, such as storms, rough seas, and tsunamis.
- **Ocean-State Forecasting**: Developing ocean-state forecasting capabilities that can predict ocean currents, waves, and other marine conditions that affect fisheries.

- **Fishery-Specific Meteorological Products**: Creating fishery-specific meteorological products, such as fishery weather forecasts, fishing area advisories, and fishery-related climate outlooks.
- **Capacity Building and Training**: Providing training and capacity-building programs for fishermen, extension workers, and other stakeholders to effectively use meteorological advisories and make informed decisions.
- **Collaboration and Partnerships**: Fostering collaboration and partnerships between meteorological agencies, fisheries departments, research institutions, and other stakeholders to improve the accuracy and relevance of meteorological advisories for the fisheries sector.
- **Leveraging Emerging Technologies**: Utilizing emerging technologies, such as artificial intelligence, machine learning, and big data analytics, to improve the accuracy and timeliness of meteorological advisories for the fisheries sector.
- **Impact-Based Forecasting**: Developing impact-based forecasting approaches that provide information on the potential impacts of weather and climate variability on fisheries, enabling more effective decision-making and risk management.

11.8.1 Benefits of Improved Meteorological Advisories

Improved meteorological advisories can benefit the fisheries sector in several ways, including:

- **Enhanced Safety**: Accurate weather forecasts can help reduce the risk of accidents and fatalities at sea.
- **Increased Productivity**: By planning fishing activities according to weather forecasts, fishermen can reduce losses and increase productivity.
- **Better Decision-Making**: Meteorological advisories can inform decision-making on fishing activities, fish processing, and marketing, enabling more effective risk management and business planning.
- **Climate Resilience**: By providing climate information and advisories, the fisheries sector can better prepare for and respond to climate variability and change.

11.8.2 Challenges and Opportunities

While there are many benefits to improving meteorological advisories for the fisheries sector, there are also challenges and opportunities to consider, such as:

- **Data Quality and Availability**: Ensuring the quality and availability of meteorological data is critical for accurate forecasting.
- **Stakeholder Engagement**: Engaging with stakeholders, including fishermen, extension workers, and policymakers, is essential for developing effective meteorological advisories.
- **Technology Transfer**: Transferring emerging technologies and research findings to operational settings can be a challenge, requiring investment in infrastructure and capacity building.
- **Sustainability and Funding**: Ensuring the sustainability and funding of meteorological advisory services is crucial for their long-term effectiveness.

By addressing these challenges and opportunities, new initiatives can enhance the effectiveness of operational meteorological advisories for the fisheries sector, supporting more informed decision-making and improved resilience to weather and climate variability.

Weather Forecasting

1. **What is weather forecasting?**
   a) Predicting earthquakes   b) Predicting agricultural yield
   c) Predicting atmospheric conditions   d) Predicting ocean currents
2. **Which of the following is NOT a weather element used in forecasting?**
   a) Temperature   b) Wind
   c) Humidity   d) Soil fertility
3. **Who is considered the father of modern weather forecasting?**
   a) Galileo Galilei   b) Sir Isaac Newton
   c) Lewis Fry Richardson   d) Carl-Gustaf Rossby
4. **Which instrument is used to measure atmospheric pressure?**
   a) Thermometer   b) Anemometer
   c) Barometer   d) Hygrometer
5. **A weather forecast valid for the next 24–48 hours is called:**
   a) Short-range forecast   b) Nowcast
   c) Medium-range forecast   d) Long-range forecast
6. **What is the role of satellites in weather forecasting?**
   a) Measuring soil health   b) Observing celestial bodies
   c) Monitoring cloud patterns and temperature   d) Predicting earthquakes

7. Weather models are based on:

a) Literary theories
b) Empirical studies only
c) Mathematical and physical principles
d) Random guesses

8. Numerical Weather Prediction (NWP) uses:

a) Hand-written charts
b) Human judgment alone
c) Supercomputers to solve equations
d) Radio broadcasts

9. Which Indian agency is responsible for official weather forecasts?

a) ICAR
b) IMD
c) ISRO
d) CSIR

10. Which of the following is a benefit of weather forecasting in fisheries?

a) Forecasting consumer demand
b) Preventing water pollution
c) Scheduling fishing trips and protecting assets
d) Enhancing fertilizer use

11. What does a Doppler radar primarily detect?

a) Salinity
b) Cloud height
c) Wind movement and precipitation
d) UV radiation

12. Agrometeorological forecasts support:

a) Industrial planning
b) Stock market predictions
c) Agricultural and fisheries operations
d) Legal procedures

13. What is the major difference between weather and climate?

a) Duration
b) Location
c) Severity
d) Accuracy

14. Which type of forecast helps in planning long-term agricultural strategy?

a) Nowcast
b) Short-range
c) Seasonal
d) Ultraviolet

15. The abbreviation 'GFS' in forecasting stands for:

a) General Forecasting System
b) Global Food Survey
c) Global Forecast System
d) Government Forecast Services

16. Nowcasting is useful for:

a) Long-term crop planning
b) Predicting seasonal drought
c) Immediate and local-scale hazards
d) Designing infrastructure

17. Which parameter is NOT typically included in a marine weather forecast?

a) Sea surface temperature
b) Wave height
c) Ocean salinity
d) Wind speed

18. Which of the following data is crucial for accurate weather forecasting?

a) Historical maps
b) Real-time atmospheric observations
c) Satellite TV signals
d) Seismic data

19. The accuracy of forecasts usually decreases with:

a) Improved technology
b) Increased data availability
c) Longer forecast period
d) Use of supercomputers

20. Meteorological observatories are located to:

a) Control temperatures
b) Measure radiation from stars
c) Collect regional weather data
d) Monitor fisheries

21. Which of these is used to disseminate agromet advisories to farmers?

a) FM radio only
b) TV only
c) SMS, apps, radio, and KVKs
d) Email to meteorologists only

22. Weather symbols on maps help represent:

a) Soil chemistry
b) Human population
c) Meteorological conditions
d) Legal boundaries

23. Probability forecasts express:

a) Guarantee of weather events
b) Certainty of predictions
c) Likelihood based on statistical models
d) Opinion-based judgment

24. A synoptic chart shows:

a) Economic trends
b) Soil moisture
c) Instantaneous atmospheric conditions
d) Air pollution

25. Which term refers to graphical weather forecasts on maps?

a) Infographics
b) Weather drawings
c) Weather charts
d) Climate diagrams

26. What is an isobar on a weather map?

a) Line joining points of equal rainfall
b) Line joining points of equal pressure
c) Line joining points of equal temperature
d) Line joining pressure centers

27. Which color in forecast maps usually indicates high rainfall?

a) Red
b) Blue
c) Green
d) Yellow

28. Why is feedback from farmers important in agromet forecasting?

a) For climate modeling
b) To test GPS instruments
c) For improving the relevance and accuracy of advisories
d) For agricultural taxes

29. The term "user-centric forecasting" refers to forecasts that:

a) Are purely scientific
b) Focus on economic forecasting
c) Cater to specific needs of end users like farmers and fishers
d) Target urban users only

30. Early warning systems aim to:

a) Scare people
b) Predict crime rates
c) Prevent disaster losses
d) Forecast stock prices

31. Which of the following weather conditions is most hazardous for fishing operations?

a) Clear skies
b) Calm winds
c) Sudden squalls
d) Low humidity

32. What is the impact of temperature forecasting on aquaculture?

a) No impact
b) Helps determine harvest timing and stress management
c) Only affects feed composition
d) Influences soil treatment

33. Why is wave height forecasting critical for fishers?

a) To avoid catching small fish
b) To prevent net loss and boat capsizing
c) To measure salinity
d) For plankton detection

34. Which forecast product is particularly relevant for coastal aquaculture farmers?

a) Soil pH forecast
b) Rainfall anomaly chart
c) Tide and surge forecast
d) Trade balance chart

35. Sudden temperature drops in forecast alerts help fish farmers to:

a) Increase feeding
b) Apply pesticides
c) Reduce stocking density or provide aeration
d) Add fertilizers

36. Rainfall forecasts help inland fish ponds avoid:

a) Excess water and overflow damage
b) Fish aging
c) Feed contamination from sun exposure
d) Aquatic insect breeding

37. Cloudy weather forecasts are crucial for:

a) Fertilizer application
b) Feeding management in pond culture
c) Transportation planning
d) Broodstock conditioning

38. Forecasting wind speed is essential to predict:

a) Oxygen deficiency in water
b) Light penetration
c) Pond seepage
d) Algal type

39. Which combination of forecasts is most useful to fishers at sea?

a) UV index, barometric pressure, soil pH
b) Wind speed, wave height, rainfall probability
c) Evapotranspiration, tide level, temperature
d) Solar angle, turbidity, dew point

40. What is the primary benefit of marine weather bulletins for fishers?

a) Advertise products
b) Plan leisure trips
c) Ensure safe navigation and timely returns
d) Inform about fish types

41. Which emerging technology enhances precision in weather forecasting?

a) Morse code
b) Manual drawing tools
c) Artificial Intelligence and Machine Learning
d) Typewriters

42. What is the expected outcome of better localized forecasts?

a) Decreased information overload
b) Improved safety and productivity at the village level
c) Lower communication costs
d) Less data generation

43. Which of the following is a challenge in delivering accurate forecasts to remote fishery communities?

a) High literacy rates
b) Lack of interest in weather
c) Poor connectivity and awareness
d) Too many satellites

44. Which satellite program contributes significantly to Indian weather forecasting?

a) Kalpana and INSAT series
b) Apollo mission
c) Chandrayaan
d) Tesla Starlink

45. High-resolution models help forecast:

a) Large-scale economic policy
b) Urban governance
c) Local extreme weather events like thunderstorms
d) Political outcomes

46. IMD's 'Mausam' mobile app is designed to:

a) Locate restaurants
b) Provide satellite TV
c) Share weather updates and forecasts with citizens
d) Monitor wildlife

47. Which of the following is a community-level forecasting initiative?

a) Global Positioning System
b) Fisherman Weather School
c) Cryptocurrency mining
d) Monsoon onset ritual

48. In forecasting, 'bias correction' refers to:

a) Political neutrality
b) Adjusting model output to reduce systematic errors
c) Ignoring user feedback
d) Removing irrelevant weather elements

49. Weather-based insurance depends on accurate forecasting of:

a) Political changes
b) Medical conditions
c) Rainfall, drought, and wind events
d) Fish behavior

50. In the context of fisheries, weather forecasting ultimately aims to:

a) Promote luxury tourism
b) Improve post-harvest marketing
c) Minimize risks and optimize operations
d) Maximize tax revenue

Answer Key (s)

1	c	2	d	3	c	4	c	5	a	6	c	7	c	8	c	9	b	10	c
11	c	12	c	13	a	14	c	15	c	16	c	17	c	18	b	19	c	20	c
21	c	22	c	23	c	24	c	25	d	26	b	27	a	28	c	29	c	30	c
31	c	32	b	33	b	34	c	35	c	36	a	37	b	38	a	39	b	40	c
41	c	42	b	43	c	44	a	45	c	46	c	47	b	48	b	49	c	50	c

12

Geography

Geography plays a vital role in understanding the complex dynamics of fisheries, encompassing both marine and inland ecosystems. The spatial distribution of fish populations, habitats, and environmental factors are all critical components that influence the productivity and sustainability of fisheries. By examining the geographical aspects of fisheries, researchers and managers can gain valuable insights into the interactions between fish populations, their environment, and human activities.

12.1 Importance of Geography in Fisheries

Geography is essential for understanding the spatial dynamics of fisheries, including:

- **Habitat mapping**: Identifying and mapping fish habitats, such as coral reefs, estuaries, and wetlands, is crucial for managing fisheries sustainably.
- **Fish migration patterns**: Understanding the migration patterns of fish populations helps researchers and managers develop effective conservation and management strategies.
- **Environmental factors**: Geography helps researchers understand the impact of environmental factors, such as climate change, ocean acidification, and water quality, on fish populations and ecosystems.
- **Human activities**: Geography also helps researchers understand the impact of human activities, such as fishing, coastal development, and pollution, on fisheries and ecosystems.

12.1.1 Applications of Geography in Marine and Inland Fisheries

Geography has numerous applications in both marine and inland fisheries, including:

- **Fisheries management**: Geography informs fisheries management decisions, such as identifying areas for protection, developing sustainable fishing practices, and monitoring fish populations.

- **Conservation planning**: Geography helps researchers identify areas of high conservation value and develop effective conservation plans for protected areas.
- **Sustainable development**: Geography informs sustainable development strategies, such as identifying areas suitable for aquaculture development and minimizing the impacts of coastal development on fisheries and ecosystems.
- **Climate change adaptation**: Geography helps researchers understand the impacts of climate change on fisheries and ecosystems, and develop effective adaptation strategies.

12.1.2 Overview

This chapter will explore the importance of geography in fisheries, highlighting its applications in both marine and inland ecosystems. We will examine the role of geography in understanding fish populations, habitats, and environmental factors, as well as its applications in fisheries management, conservation planning, sustainable development, and climate change adaptation. By understanding the geographical aspects of fisheries, researchers and managers can develop more effective strategies for managing fisheries sustainably and promoting the long-term health of these critical ecosystems.

12.2 The Shape, Size, and Structure of the Earth

The Earth is an intricate and dynamic planet, and understanding its shape, size, and structure is vital for various fields of study, including geology, geography, and environmental science (Fig.12.1).

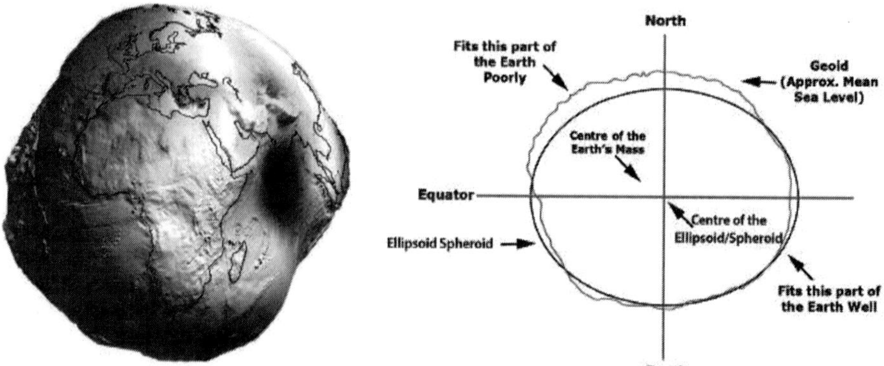

Fig. 12.1: Geoid shape of earth

12.2.1 Shape of the Earth

The Earth is an oblate spheroid, meaning that it is slightly flattened at the poles and bulging at the equator. This shape is a result of the planet's rotation, which generates centrifugal force that pushes the Earth's mass away from the axis of rotation. The Earth's shape is also influenced by its slightly ellipsoidal shape, which is caused by the planet's rotation and the movement of tectonic plates.

12.2.2 Size of the Earth

The Earth's size is remarkable, with a diameter of approximately 12,742 kilometers (7,918 miles) at the equator. The Earth's circumference at the equator is approximately 40,075 kilometers (24,901 miles). The Earth's size and shape play a crucial role in understanding various geological and environmental processes, including the planet's climate, ocean currents, and tectonic activity.

12.2.3 Structure of the Earth

The Earth is composed of several distinct layers, each with its own unique characteristics and properties. The main layers of the Earth are:

- **Crust**: The crust is the outermost layer of the Earth, ranging in thickness from 5-70 kilometers (3-44 miles). The crust is composed of rocks and minerals and is broken into several large plates that float on the more fluid layers below.
- **Mantle**: The mantle is the layer beneath the crust, extending from a depth of about 35 kilometers (22 miles) to 2,900 kilometers (1,800 miles). The mantle is composed of hot, viscous rock that can flow over long periods of time.
- **Outer Core**: The outer core is a liquid layer of iron and nickel, extending from a depth of about 2,250 kilometers (1,398 miles) to 5,150 kilometers (3,200 miles). The outer core is responsible for generating the Earth's magnetic field.
- **Inner Core**: The inner core is a solid, iron-nickel alloy at the center of the Earth, with a radius of about 1,220 kilometers (760 miles). The inner core is the hottest part of the Earth, with temperatures reaching around 5,000°C (9,000°F).

12.2.4 Key Features of the Earth's Structure

The Earth's structure has several key features that play a crucial role in shaping our planet. These include:

- **Tectonic plates**: The Earth's crust is broken into several large plates that move relative to each other, resulting in geological activity such as earthquakes and volcanic eruptions.
- **Mountain ranges**: The Earth's surface is home to several major mountain ranges, including the Himalayas and the Andes, which are formed by the movement of tectonic plates.
- **Ocean basins**: The Earth's surface is also home to several large ocean basins, including the Pacific and Atlantic Oceans, which play a crucial role in the planet's climate and ecosystems.

12.2.5 Importance of Understanding the Earth's Shape, Size, and Structure

Understanding the Earth's shape, size, and structure is essential for various fields of study, including:

- **Geology**: Understanding the Earth's structure and composition is crucial for understanding geological processes, such as plate tectonics and earthquakes.
- **Environmental science**: Understanding the Earth's shape, size, and structure is essential for understanding environmental processes, such as climate change and ocean currents.
- **Navigation and mapping**: Understanding the Earth's shape and size is crucial for navigation and mapping, including GPS technology and cartography.

12.2.6 Applications of Earth's Shape, Size, and Structure

The Earth's shape, size, and structure have numerous applications in various fields, including:

- **Climate modeling**: Understanding the Earth's shape and size is crucial for climate modeling and predicting future climate changes.
- **Natural resource management**: Understanding the Earth's structure and composition is essential for managing natural resources, such as minerals and water.
- **Disaster risk reduction**: Understanding the Earth's structure and geological processes is crucial for disaster risk reduction and mitigation.

In conclusion, the Earth's shape, size, and structure are complex and fascinating topics that are essential for understanding our planet and its many geological and environmental processes. By studying the Earth's shape, size, and structure, we can gain a deeper understanding of the planet's dynamics and better manage natural resources, predict natural disasters, and mitigate the impacts of climate change.

12.3 Understanding Latitude, Longitude, and Great Circles

12.3.1 Latitude

Latitude is a fundamental concept in geography that refers to the angular distance of a place north or south of the Earth's equator. It's measured in degrees, minutes, and seconds, with 0° at the equator and 90° at the poles. Latitude plays a crucial role in determining the climate, temperature, and weather patterns of a region (Fig. 12.2).

- **Key Features of Latitude**
  - **Measurement**: Latitude is measured in degrees, minutes, and seconds.
  - **Range**: Latitude ranges from 0° to 90° north (N) and 0° to 90° south (S).
  - **Parallels**: Lines of latitude are also known as parallels, which are parallel to the equator.
  - **Climate**: Latitude influences the climate of a region, with temperatures generally decreasing as you move towards the poles.

12.3.2 Longitude

Longitude is another essential concept in geography that refers to the angular distance of a place east or west of the prime meridian in Greenwich, England. It's also measured in degrees, minutes, and seconds, with 0° at the prime meridian and 180° east and west. Longitude is critical for determining time zones and navigation routes.

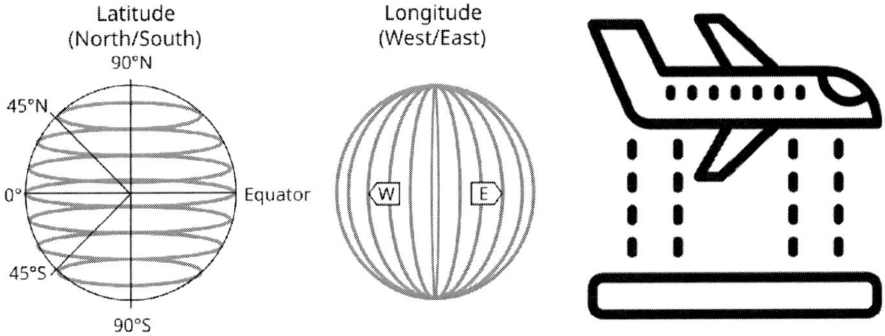

Fig. 12.2: Geo-coordinates

- **Key Features of Longitude**
  - **Measurement**: Longitude is measured in degrees, minutes, and seconds.
  - **Range**: Longitude ranges from 0° to 180° east (E) and 0° to 180° west (W).
  - **Meridians**: Lines of longitude are also known as meridians, which converge at the poles.
  - **Time Zones**: Longitude influences time zones, with each 15° of longitude representing a one-hour difference.

12.3.3 Great Circles

A great circle is an imaginary circle on the surface of a sphere (like the Earth) that is centered on the sphere's center. It's the shortest path between two points on the surface of a sphere, making it a vital concept in navigation.

- **Key Features of Great Circles**
  - **Shortest Distance**: Great circles represent the shortest distance between two points on the surface of a sphere.
  - **Navigation**: Great circles are used in navigation to find the shortest distance between two points on the Earth's surface.
  - **Formula**: The great circle formula calculates the distance between two points given their latitudes and longitudes.
  - **Applications**: Great circles have applications in aviation, maritime, and geography.

12.3.4 Calculating Great Circle Distance

The great circle distance formula is given by:

d = r * cos^(-1) [cos(a) * cos(b) * cos(x-y) + sin(a) * sin(b)]

where:

- d is the distance between the two points
- r is the radius of the Earth (approximately 6,371 km)
- a and b are the latitudes of the two points
- x and y are the longitudes of the two points

12.3.5 Applications of Latitude, Longitude, and Great Circles

Latitude, longitude, and great circles have numerous applications in various fields, including:

- **Navigation**: Understanding latitude, longitude, and great circles is essential for navigation in aviation and maritime.
- **Geography**: Latitude, longitude, and great circles are used to study the Earth's surface and understand spatial relationships between different locations.
- **Cartography**: Latitude, longitude, and great circles are used in map projections to represent the Earth's surface accurately.
- **Climate Modeling**: Latitude and longitude are used in climate modeling to understand and predict climate patterns.

In conclusion, latitude, longitude, and great circles are fundamental concepts in geography that have numerous applications in various fields. Understanding these concepts is essential for navigation, geography, cartography, and climate modeling

12.4 Model Globe

The term "model globe" can refer to a few different concepts, primarily within Earth science and related fields (Fig. 12.3). Here are some of the key interpretations:

Fig. 12.3: Model globe

12.4.1 Physical Model of the Earth

- This is likely the most common understanding of a "model globe." It refers to a spherical representation of the Earth, depicting landmasses, oceans, and often features like latitude and longitude lines, and sometimes even topography.
- **Purpose:** Physical model globes are used for educational purposes, for visualizing the spherical shape of the Earth, understanding geographical relationships, and sometimes for decorative purposes.
- **Significance:** They provide a tangible and accurate (in terms of shape) representation of our planet, unlike flat maps which inherently involve distortions.

12.4.2 Global-Scale Computational Models (as discussed in the fisheries context)

- In more scientific contexts, "model globe" can metaphorically refer to complex computer simulations that represent the Earth's systems on a global scale. These are not physical objects but rather sophisticated software.
- **Types:** These include:
  - **Global Climate Models (GCMs):** These models simulate the Earth's climate system, including the atmosphere, oceans, land surface, and ice. They are used to understand past climate, simulate the present, and project future climate change.
  - **Earth System Models (ESMs):** These are more comprehensive than GCMs, as they also include representations of the Earth's biogeochemical cycles (e.g., the carbon cycle) and interactions with the biosphere.
- **Purpose:** To understand the complex interactions within the Earth system, to make predictions about future environmental changes, and to assess the impacts of human activities.
- **Significance:** These "model globes" are crucial tools for climate science, environmental research, and informing policy decisions. They allow scientists to conduct experiments that are impossible in the real world, such as simulating the effects of increased greenhouse gases.

12.4.3 Key Differences

- The first type is a **physical object**, while the second is a **computational tool**.

- The first is primarily for **visualization and basic understanding**, while the second is for **complex scientific analysis and prediction**.

Given that the initial query was in the context of fisheries, the "model globe" mentioned there aligns with the second interpretation – global-scale computational models that include ocean components relevant to marine life and fisheries. However, the term itself can also simply refer to a physical globe of the Earth.

Okay, let's dive into map projections with even more detail, thinking about the "how" and "why" behind them.

Imagine the Earth as a perfectly smooth, transparent ball with all the continents and lines of latitude and longitude painted on it. Our goal is to take this 3D information and put it onto a 2D flat surface – a map.

12.4.4 The Challenge: Squashing a Sphere

The fundamental problem is that a sphere is curved in two directions, while a flat map is only curved in zero directions. To make something curved flat, you have to stretch, compress, or tear it. Map projections are the mathematical formulas we use to decide *how* to do this "flattening" in a systematic way.

12.5 The "Projection" Idea

Think of shining a light from the center of our transparent globe. The shadows of the continents and the grid lines would fall onto a surface we've placed around the globe. The shape of this surface (cylinder, cone, or a flat plane) determines the *type* of projection. The Three Main "Developable Surfaces": are:

These are the simple geometric shapes that can be laid flat without stretching or tearing.

12.5.1 Cylinder

- Imagine wrapping a cylinder of paper around the globe. The projection happens onto this cylinder.
- When you unroll the cylinder, you get a rectangular map.
- **Characteristics:** Meridians (lines of longitude) become straight, vertical, and equally spaced. Parallels (lines of latitude) become straight, horizontal, and equally spaced (though their spacing might change depending on the specific cylindrical projection).

- **Distortion:** Typically, distortion is minimal near the line of tangency (where the cylinder touches the globe, often the Equator). Distortion of area and shape increases as you move away from this line, towards the poles.
- **Examples**
  - **Mercator:** This is a *conformal* projection, meaning it preserves angles locally. This makes it great for navigation because lines of constant bearing (rhumb lines) appear as straight lines. However, it severely distorts area, making landmasses at high latitudes (like Greenland and Antarctica) appear much larger than they are in reality. The parallels of latitude are spaced farther apart as you move towards the poles.
  - **Equirectangular (or Plate Carrée):** This is one of the simplest. Meridians and parallels are straight and equally spaced. However, it significantly distorts both shape and area, especially away from the standard parallels (often the Equator).

12.5.2 Cone

- Imagine placing a cone over the globe, with its tip usually above one of the poles. The projection happens onto this cone.
- When you unroll the cone, you get a fan-shaped map.
- **Characteristics:** Meridians are straight lines radiating from the apex (tip) of the cone, and parallels are concentric arcs (parts of circles) centered at the apex.
- **Distortion:** Distortion is usually least along the *standard parallels* – the lines of latitude where the cone touches the globe. Distortion increases as you move away from these standard parallels. Conic projections are often good for mapping mid-latitude regions.
- **Examples**
  - **Lambert Conformal Conic:** This projection preserves angles and is often used for mapping regions that are longer east-west, like the contiguous United States. It typically uses two standard parallels to minimize distortion across the mapped area.
  - **Albers Equal-Area Conic:** This projection preserves area, making it useful for thematic maps where the size of regions needs to be accurately represented. It also usually uses two standard parallels.

12.5.3 Plane (Azimuthal or Planar)

- Imagine touching a flat piece of paper to the globe at a single point. The projection happens onto this plane.

- The resulting maps are circular.
- **Characteristics:** Straight lines radiating from the center point of the projection represent great circles (the shortest distance between two points on a sphere). Distances and/or directions from the center point might be preserved, depending on the specific azimuthal projection.
- **Distortion:** Distortion generally increases as you move away from the central point.
- **Examples**
  - **Azimuthal Equidistant:** This projection preserves distances from the central point. It's useful for showing airline distances from a specific city.
  - **Gnomonic:** This projection shows all great circles as straight lines, important for navigation by air or sea over long distances. However, it severely distorts shape and area away from the center.
  - **Orthographic:** This projection creates the illusion of looking at the Earth from space. Areas near the edges appear compressed.

12.5.4 Beyond the Basics: Compromise and other Projections

Not all projections neatly fit into the cylinder, cone, or plane categories.

- **Pseudocylindrical Projections:** These try to combine some of the best aspects of cylindrical and other projections. They often have straight parallels and a central meridian, but other meridians are curved. Examples include the Mollweide and Sinusoidal projections, which are often used for world maps showing area accurately, at the cost of some shape distortion.
- **Compromise Projections:** These projections don't perfectly preserve any single property (like area, shape, or distance) but aim for a balance of distortions to make the map look "right." The **Robinson projection** is a popular example for world maps.

12.5.5 The Trade-Offs

The key takeaway is that every map projection involves trade-offs. You can't have a flat map that perfectly represents all aspects of the Earth simultaneously. Cartographers (mapmakers) choose the projection that best suits the map's purpose.

- **Conformal (shape-preserving):** Good for navigation and showing local shapes accurately.

- **Equal-area (area-preserving):** Essential for thematic maps where the size of regions is important.
- **Equidistant (distance-preserving):** Useful for showing accurate distances from a central point.
- **True direction:** Some projections accurately show directions from a central point.

Understanding map projections helps us to critically look at maps and realize that the way the world is presented is not the only way it *could* be presented. Different projections tell different visual stories and are useful for different things.

12.6 Cartography: The Art, Science, and Technology of Making Maps

Cartography is the discipline that deals with the **conception, production, dissemination, and study of maps** (Fig. 12.4). It's more than just drawing lines on paper; it's a blend of:

- **Art:** Involves design, aesthetics, and visual communication to make maps clear, engaging, and effective.
- **Science:** Relies on surveying, geodesy, photogrammetry, and increasingly, computer science and GIS (Geographic Information Systems) to ensure accuracy and manage spatial data.
- **Technology:** Utilizes various tools from traditional drawing instruments to sophisticated software for data processing, analysis, and map creation.

Essentially, cartography is about representing the spatial relationships of the real world (or even imagined worlds) in a visual and understandable format.

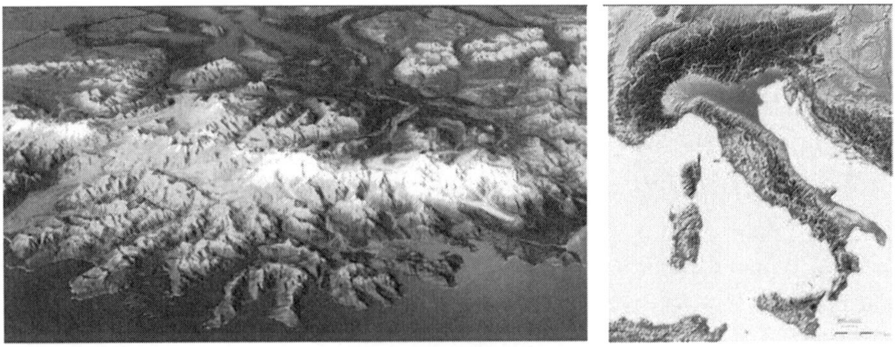

Fig. 12.4: Cartography

12.6.1 Applications in Fisheries

Cartography, particularly through the lens of modern GIS and remote sensing, has become an invaluable tool in various aspects of fisheries science and management. Here's how:

- **Habitat Mapping and Conservation**
  - **Identifying critical habitats:** Mapping spawning grounds, nursery areas, feeding zones, coral reefs, seagrass beds, and other essential fish habitats.
  - **Monitoring habitat health:** Tracking changes in these habitats due to natural events or human activities (e.g., pollution, coastal development, climate change).
  - **Designing Marine Protected Areas (MPAs):** Using spatial data to identify optimal locations and boundaries for conservation efforts.
- **Fisheries Management**
  - **Mapping fishing grounds:** Identifying and characterizing areas where fishing occurs, including the distribution of different fishing activities.
  - **Monitoring fishing effort:** Tracking the spatial distribution and intensity of fishing fleets using vessel monitoring systems (VMS).
  - **Stock assessment:** Combining catch data with spatial information to understand the distribution and abundance of fish stocks.
  - **Spatial management measures:** Defining closed areas, seasonal closures, and other geographically specific regulations.
  - **Resolving conflicts:** Mapping the spatial overlap between different user groups (e.g., commercial, recreational fishers) to help manage potential conflicts.
- **Monitoring and Predicting Fish Movements**
  - **Tracking tagged fish:** Using telemetry data linked to maps to understand migration patterns and habitat use.
  - **Modeling species distribution:** Predicting where fish species are likely to be found based on environmental factors (e.g., temperature, depth, salinity) mapped cartographically.
- **Ecosystem Modeling and Productivity Assessments**
  - **Mapping environmental variables:** Visualizing data on sea surface temperature, chlorophyll concentration, ocean currents, and other factors that influence marine productivity.

    - **Analyzing spatial relationships:** Understanding how environmental conditions relate to the distribution and abundance of fish and other marine organisms.

- **Climate Change and Fisheries**
    - **Mapping potential impacts:** Visualizing how changes in sea level, temperature, and ocean chemistry might affect fish habitats and distributions.
    - **Identifying vulnerable areas:** Pinpointing regions and species most at risk from climate change impacts.

- **Aquaculture Site Selection**
    - **Identifying suitable locations:** Mapping areas based on water quality, accessibility, environmental sensitivity, and other relevant criteria for aquaculture development.

In essence, cartography provides the **spatial context** necessary to understand and manage fisheries effectively. By visualizing and analyzing geographically referenced data, it enables better decision-making for sustainable resource use and conservation.

12.6.2 Landscape: The Visible Features of an Area and Its Relevance to Fisheries

In its most general sense, a **landscape** refers to the **visible features of an area of land**. This includes:

- **Landforms:** Mountains, hills, valleys, plains.
- **Water bodies:** Rivers, lakes, ponds, coastlines, seas.
- **Living elements:** Vegetation (forests, grasslands), and even animals to some extent.
- **Human elements:** Buildings, roads, agricultural fields, and other modifications by people.

Increasingly, especially in the field of **landscape ecology**, the concept goes beyond just the visual. It considers a landscape as a **mosaic of interacting ecosystems** that are repeated in a similar form across an area. This emphasizes the spatial arrangement of different habitat types and how they connect and influence ecological processes.

12.6.3 Relevance to Fisheries

While we often think of fisheries as being solely about the aquatic environment, the surrounding terrestrial **landscape** has a profound and often overlooked influence on the health, productivity, and management of fisheries. Here's how:

- **Water Quality**
  - **Runoff:** The landscape dictates how water flows into rivers, lakes, and coastal areas. Land use (e.g., agriculture, urbanization, forestry) in the watershed directly affects the amount and type of pollutants (sediments, nutrients, pesticides) that enter aquatic ecosystems, impacting water quality crucial for fish survival and reproduction.
  - **Riparian Zones:** Vegetation along riverbanks and lake shores (part of the landscape) acts as a natural filter, preventing pollutants from entering the water and providing shade that regulates water temperature.
- **Habitat Structure and Connectivity**
  - **Physical Habitat:** The geology and topography of the landscape influence the physical structure of aquatic habitats (e.g., the shape and flow of rivers, the depth and substrate of lakes).
  - **Connectivity:** The landscape can either facilitate or hinder the movement of fish. Dams, roads, and other landscape features can fragment aquatic habitats, preventing migration for spawning or access to different feeding areas. Conversely, natural corridors within the landscape can support connectivity.
- **Nutrient Cycling and Energy Flow**
  - **Terrestrial Inputs:** Organic matter and nutrients from the surrounding terrestrial landscape (e.g., leaf litter) can be important sources of energy and nutrients for aquatic food webs, supporting fish populations.
- **Climate Regulation**
  - Large-scale landscape features like forests can influence local and regional climate, which in turn affects water temperature and precipitation patterns vital for aquatic ecosystems and fish.
- **Human Interactions:**
  - The way humans use the landscape (e.g., fishing practices, coastal development, agriculture) directly impacts fisheries. Understanding the spatial distribution of these activities within the broader landscape is crucial for management.

In essence, a healthy and well-functioning terrestrial landscape often translates to healthier and more productive aquatic ecosystems, which directly benefits fisheries. Ignoring the landscape context can lead to incomplete or ineffective fisheries management strategies.

Landscape ecology provides a framework for understanding these interactions between terrestrial and aquatic systems, emphasizing the importance of spatial patterns and ecological processes across different scales. Applying a landscape perspective to fisheries management can lead to more holistic and sustainable approaches.

Geography

1. **Which of the following is the innermost layer of the Earth?**
   a) Mantle b) Crust
   c) Inner core d) Outer core
2. **The movement of tectonic plates is driven by:**
   a) Solar radiation b) Ocean currents
   c) Convection currents in the mantle d) Earth's magnetic field
3. **Which type of rock is formed from cooling of molten magma?**
   a) Sedimentary b) Metamorphic
   c) Igneous d) Fossiliferous
4. **The Himalayas were formed by the collision of which two plates?**
   a) Indo-Australian and Eurasian b) Pacific and North American
   c) African and Arabian d) Nazca and South American
5. **Which river forms the largest delta in the world?**
   a) Amazon b) Ganges-Brahmaputra
   c) Mississippi d) Nile
6. **Which desert is the largest in the world by area?**
   a) Gobi b) Sahara
   c) Thar d) Kalahari
7. **Which of the following is a depositional feature of a river?**
   a) Waterfall b) Canyon
   c) Delta d) Gorge
8. **Which layer of the atmosphere contains the ozone layer?**
   a) Troposphere b) Stratosphere
   c) Mesosphere d) Thermosphere

9. Which mountain range is the oldest in India?

a) Himalayas
b) Aravalli
c) Western Ghats
d) Satpura

10. The longest river in the world is:

a) Amazon
b) Ganga
c) Yangtze
d) Nile

11. Earthquakes are measured using:

a) Thermometer
b) Anemometer
c) Seismograph
d) Barometer

12. Which ocean is the deepest in the world?

a) Indian
b) Atlantic
c) Arctic
d) Pacific

13. Which of the following is a metamorphic rock?

a) Granite
b) Basalt
c) Slate
d) Limestone

14. The Deccan Plateau is made up of:

a) Sedimentary rocks
b) Igneous rocks
c) Metamorphic rocks
d) Alluvial soil

15. Which phenomenon causes day and night?

a) Revolution of Earth
b) Rotation of Earth
c) Tilt of Earth
d) Precession

16. The Tropic of Cancer passes through how many Indian states?

a) 5
b) 6
c) 7
d) 8

17. The term "Ring of Fire" is associated with:

a) Earthquakes in the Atlantic
b) Volcanic zone around the Pacific
c) Deserts in Africa
d) Drought-prone areas of Australia

18. The point on the Earth's surface directly above the earthquake origin is called:

a) Seismic point
b) Focus
c) Epicenter
d) Fault line

19. Which river is known as the "Sorrow of Bihar"?

a) Ganga
b) Ghaghra
c) Kosi
d) Son

20. Which current is warm?

a) Labrador
b) Humboldt
c) Canary
d) Gulf Stream

21. Which country has the highest population density?

a) USA
b) China
c) India
d) Bangladesh

22. Which factor affects population distribution the most?

a) Soil fertility
b) Altitude
c) Climate
d) Culture

23. The term 'GDP' stands for:

a) Gross Daily Product
b) General Domestic Price
c) Gross Domestic Product
d) Global Domestic Population

24. Which crop is known as a commercial crop in India?

a) Rice
b) Wheat
c) Cotton
d) Maize

25. Which sector contributes most to India's GDP?

a) Agriculture
b) Industry
c) Services
d) Mining

26. Which of the following cities is known for its IT industry in India?

a) Chennai
b) Mumbai
c) Bengaluru
d) Delhi

27. Which country is the largest producer of wheat?

a) USA
b) China
c) India
d) Russia

28. What is urbanization?

a) Growth of crops
b) Migration of animals
c) Growth of cities
d) Climate modification

29. Which state in India has the highest forest cover by area?

a) Assam
b) Madhya Pradesh
c) Kerala
d) Odisha

30. Which resource is renewable?

a) Petroleum
b) Coal
c) Solar energy
d) Natural gas

31. Which Indian port is the largest by volume of cargo handled?

a) Kandla
b) Mumbai
c) Visakhapatnam
d) Paradip

32. Which is the most urbanized state in India?

a) Maharashtra b) Tamil Nadu

c) Goa d) Kerala

33. Which mineral is used in nuclear power generation?

a) Iron b) Copper

c) Uranium d) Manganese

34. The Golden Quadrilateral connects:

a) Four rivers b) Four major cities

c) Four ports d) Four states

35. India's largest coal reserves are found in:

a) Odisha b) Jharkhand

c) Madhya Pradesh d) Chhattisgarh

36. The major cause of acid rain is:

a) CO_2 emissions b) Methane

c) Sulphur and nitrogen oxides d) Water vapor

37. Which gas is primarily responsible for global warming?

a) Nitrogen b) Oxygen

c) Carbon dioxide d) Helium

38. Which protocol is related to the reduction of greenhouse gases?

a) Geneva Protocol b) Montreal Protocol

c) Kyoto Protocol d) Basel Convention

39. Chipko movement was aimed at:

a) Soil conservation b) Saving water

c) Forest conservation d) Saving energy

40. The ozone hole is most prominent over:

a) Asia b) Antarctica

c) Europe d) North America

41. Which branch of geography is most relevant to fisheries studies?

a) Political geography b) Economic geography

c) Physical geography d) Urban geography

42. Cartography is the science and art of:

a) Measuring ocean depth b) Drawing and interpreting maps

c) Fishing techniques d) Predicting weather

43. Topographic maps are most useful in fisheries for:

a) Understanding fish breeds
b) Identifying elevation and water bodies
c) Assessing rainfall patterns
d) Forecasting temperature

44. Which tool is commonly used in modern cartography for fisheries resource mapping?

a) Thermometer
b) Barometer
c) GIS (Geographic Information System)
d) Trawl net

45.The intersection of geography and fisheries primarily focuses on:

a) Mapping fish markets
b) Understanding how physical features affect aquatic ecosystems
c) Making fishing equipment
d) Legal aspects of fishing

46. Which of the following features is best shown on a hydrographic map?

a) Political boundaries
b) Rainfall zones
c) Depth of water bodies and seabed features
d) Soil types

47. Latitude and longitude help in:

a) Forecasting rainfall
b) Determining global positioning of water bodies
c) Observing cloud types
d) Tracking fish behavior

48. A bathymetric map is designed to represent:

a) Wind direction
b) Subsurface topography of ocean/lake floors
c) Forest cover
d) Population density

Answer: b) Subsurface topography of ocean/lake floors

49. Remote sensing contributes to fisheries by:

a) Teaching manual netting
b) Tracking fish prices
c) Monitoring aquatic vegetation and water quality
d) Analyzing soil erosion

50. What type of scale is used on maps to measure real-world distances?

a) Bar scale or graphical scale
b) Vernier scale
c) Compass scale
d) Numeric timer

Answer Key (s)

1	c	2	c	3	c	4	a	5	b	6	b	7	c	8	b	9	b	10	d
11	c	12	d	13	c	14	b	15	b	16	d	17	b	18	c	19	c	20	d
21	d	22	c	23	c	24	c	25	c	26	c	27	b	28	c	29	b	30	c
31	a	32	c	33	c	34	b	35	b	36	c	37	c	38	c	39	c	40	b
41	c	42	b	43	b	44	c	45	b	46	c	47	b	48	b	49	c	50	a

13

Effect of Climate Change on the Fisheries Sector

The world's fisheries, encompassing both marine and freshwater environments, play a critical role in global food security, livelihoods, and economies. However, these vital resources are increasingly threatened by the far-reaching consequences of climate change. This chapter delves into the multifaceted impacts of climate change on the fisheries sector, exploring the complex interplay between rising temperatures, altered weather patterns, and ocean acidification, and how these changes affect fish populations, habitats, and the communities that depend on them.

We will examine the ways in which climate change is altering the fundamental dynamics of aquatic ecosystems, leading to shifts in species distribution, disruptions in food webs, and increased vulnerability to extreme weather events. Special attention will be given to the differential impacts on various regions and fisheries, considering the unique challenges faced by small-scale fisheries and developing nations. Furthermore, the chapter will discuss the potential consequences for global fish production, economic stability, and food security, highlighting the urgent need for sustainable management practices and adaptive strategies to mitigate these adverse effects and ensure the long-term viability of the fisheries sector.

13.1 Climate Change Impacts on Marine Fisheries

Climate change is significantly altering marine ecosystems, leading to profound impacts on fisheries. These changes affect various aspects of fish biology and ecology, including fishery dynamics, migration patterns, and breeding seasons (Fig.13.1). Here's a detailed breakdown:

Fig. 13.1: Climate change impact on fisheries

13.1.1 Fishery Dynamics

- **Changes in Productivity**
  - **Primary Production:** Climate change affects the base of the marine food web by altering phytoplankton productivity. Rising temperatures and ocean acidification can disrupt phytoplankton growth, affecting the availability of food for fish. In some areas, productivity may increase, while in others, particularly in nutrient-poor regions, it may decline.
  - **Food Web Disruption:** Changes in primary production can cascade through the food web, affecting the abundance and distribution of zooplankton, which are crucial food sources for many fish species. This can lead to mismatches between predator and prey, impacting fish growth, survival, and reproductive success.
  - **Species Composition:** Climate change can cause shifts in species composition within marine ecosystems. Some species may thrive in warmer waters, while others decline. This can lead to changes in the overall productivity and composition of fish stocks, affecting the catch potential of fisheries.
- **Stock Abundance and Distribution**
  - **Recruitment:** Climate change can affect fish recruitment (the number of young fish that survive to adulthood) through various mechanisms, including changes in spawning success, larval survival, and the availability of suitable nursery habitats.
  - **Growth Rates:** Warmer temperatures can increase metabolic rates in fish, leading to faster growth in some species. However, beyond a certain threshold, increased temperatures can lead to thermal stress, reduced growth, and smaller body sizes.

- **Mortality:** Climate change can increase fish mortality rates through several pathways, including increased frequency and intensity of heatwaves, altered disease prevalence, and increased predation pressure due to changes in predator-prey interactions.

- **Extreme Events**
  - **Marine Heatwaves:** These events can cause widespread mortality of marine organisms, including fish, and disrupt entire ecosystems.
  - **Storms and Cyclones:** Increased frequency and intensity of storms can damage fishing infrastructure, disrupt fishing activities, and alter coastal habitats, affecting fish populations.

13.1.2 Migration Patterns

- Range Shifts
  - **Poleward Migration:** Many fish species are shifting their ranges towards the poles in response to warming ocean temperatures. This can lead to changes in the distribution of fish stocks, affecting where fishing can occur and potentially leading to conflicts over resources.
  - **Deepwater Migration:** Some species are also migrating to deeper, cooler waters to escape warming surface temperatures. This can make them harder to catch and change the dynamics of deep-sea ecosystems.
- Timing of Migration:
  - **Phenological Changes:** Climate change can alter the timing of fish migrations, leading to mismatches with the availability of food or suitable spawning conditions. This can affect fish survival and reproductive success.
  - **Altered Currents:** Changes in ocean currents can disrupt migratory pathways, making it difficult for fish to reach their spawning or feeding grounds.

13.1.3 Breeding Seasons

- **Spawning Timing**
  - **Temperature Dependence:** Fish spawning is often triggered by temperature cues. Climate change can cause earlier or delayed spawning, potentially leading to mismatches with the availability of food for larvae.
  - **Reproductive Success:** Changes in water temperature, salinity, and ocean acidification can directly affect the survival and development of fish eggs and larvae, impacting reproductive success.

- **Spawning Grounds**
    - **Habitat Degradation:** Climate change can degrade critical spawning habitats, such as coral reefs, mangroves, and estuaries, through ocean acidification, sea-level rise, and increased storm intensity.
    - **Changes in Water Quality:** Altered salinity levels, reduced oxygen concentrations, and increased pollution due to climate change can negatively impact spawning success.

13.2 Effect of Global Weather Parameters on Marine Fisheries

Global weather parameters play a crucial role in shaping the dynamics of marine ecosystems and, consequently, the productivity and sustainability of marine fisheries. Changes in these parameters, driven by climate change, can have profound impacts on fish populations, their habitats, and the livelihoods of people who depend on them. Here's how:

13.2.1 Key Global Weather Parameters and Their Effects

- **Sea Surface Temperature (SST)**
    - **Impact:** SST is a critical factor influencing fish metabolism, growth, distribution, and reproduction. Rising SST can cause fish to migrate to cooler waters, alter spawning seasons, and disrupt food webs.
    - **Example:** The northward shift of Atlantic cod and mackerel in response to warming waters has been observed in the North Atlantic, affecting fisheries management and international agreements.
- **Ocean Acidification**
    - **Impact:** Increased absorption of CO_2 by the ocean leads to lower pH levels, making it difficult for calcifying organisms like corals, shellfish, and plankton to build their skeletons and shells. This disrupts the base of the food web and alters habitats crucial for fish.
    - **Example:** The decline in oyster populations in the Pacific Northwest due to ocean acidification has significant economic consequences for the region's shellfish industry.
- **Sea Level Rise**
    - **Impact:** Rising sea levels can inundate coastal habitats like mangroves, salt marshes, and estuaries, which serve as important nursery grounds for many fish species. It also threatens coastal communities and fishing infrastructure.

- **Example:** In Southeast Asia, sea-level rise is projected to inundate critical mangrove ecosystems, impacting fish breeding and nursery areas and reducing coastal fish stocks.

- **Storms and Extreme Weather Events**
  - **Impact:** More frequent and intense storms can disrupt fishing activities, damage infrastructure, and alter marine habitats. They can also cause temporary or permanent shifts in fish distribution and abundance.
  - **Example:** Increased cyclone activity in the Bay of Bengal has disrupted fishing operations and damaged coastal aquaculture farms, affecting the livelihoods of fishing communities.

- **Ocean Currents and Circulation**
  - **Impact:** Changes in ocean currents can affect the distribution of nutrients, larval dispersal, and migration patterns of fish. Altered upwelling patterns, for instance, can reduce the availability of nutrients, leading to decreased primary productivity and lower fish stocks.
  - **Example:** Changes in the Gulf Stream can affect the distribution and abundance of commercially important fish species in the Northwest Atlantic, impacting fisheries in the region.

- **Precipitation Patterns**
  - **Impact:** Changes in rainfall patterns can affect river runoff, salinity levels, and nutrient input into coastal areas, influencing estuarine and nearshore habitats crucial for many fish species.
  - **Example:** Increased freshwater runoff in some Arctic regions due to melting glaciers is altering salinity levels and impacting the distribution of fish species adapted to specific salinity ranges.

13.2.2 Case Studies

- **The North Atlantic Cod Fishery**
  - **Background:** The collapse of the North Atlantic cod fishery in the early 1990s was partly attributed to overfishing, but climate change also played a role.
  - **Climate Change Impact:** Changes in water temperature and ocean currents affected cod migration, spawning, and recruitment, exacerbating the impact of overfishing and hindering the recovery of the stock.

- **The California Current System**
  - **Background:** This highly productive upwelling system supports a diverse array of fish species and valuable fisheries.
  - **Climate Change Impact:** Changes in wind patterns and ocean temperatures have altered upwelling intensity, affecting nutrient availability and primary productivity. This has led to shifts in the abundance and distribution of species like salmon, sardine, and anchovy, impacting fisheries and the marine food web.
- **Coral Reef Fisheries**
  - **Background:** Coral reefs provide critical habitat for a vast number of fish species and support important fisheries in tropical regions.
  - **Climate Change Impact:** Rising ocean temperatures and ocean acidification are causing coral bleaching and degradation, leading to a loss of habitat complexity and a decline in fish populations. This threatens the livelihoods of communities that depend on coral reef fisheries for food and income.
- **Arctic Fisheries**
  - **Background:** The Arctic is experiencing rapid warming and sea ice loss, opening up new areas for fishing but also altering the ecosystem.
  - **Climate Change Impact:** Changes in sea ice extent, temperature, and freshwater input are affecting the distribution and abundance of Arctic fish species, leading to shifts in fisheries potential and posing challenges for management and conservation.

13.3 Effect of Climate Change on Inland Fisheries

Inland fisheries, which occur in freshwater environments like rivers, lakes, and wetlands, are vital sources of food and livelihoods for millions of people worldwide. However, these fisheries are particularly vulnerable to the impacts of climate change. Here's a detailed explanation of the effects:

13.3.1 Changes in Water Temperature

- **Impacts**
  - **Direct Physiological Effects:** Fish are cold-blooded, so their body temperature and metabolic rate are directly influenced by water temperature. Rising temperatures can increase their energy needs, alter growth rates, and affect reproductive success.

- **Habitat Suitability:** Many fish species have specific temperature requirements. Warming waters can force them to migrate to find suitable habitats, potentially disrupting ecosystems and fisheries. Some species may be unable to relocate and face local extinction.
- **Increased Stress and Disease:** Higher temperatures can increase stress on fish, making them more susceptible to diseases and parasites.

- **Explanation:** As global air temperatures rise, so do the temperatures of freshwater bodies. This seemingly small change can have drastic effects on the physiology and behavior of fish. For example, in some temperate regions, warmer winters may lead to earlier spawning, which could be mismatched with food availability for newly hatched fish.

13.3.2 Altered Water Flow Regimes

- **Impacts**
  - **Droughts:** Increased evaporation and altered rainfall patterns can lead to more frequent and severe droughts, reducing water levels in rivers and lakes. This can destroy fish habitats, concentrate pollutants, and limit fish movement.
  - **Floods:** Conversely, more intense rainfall events can cause more frequent and severe floods. These can damage fish habitats, disrupt spawning, and wash fish away from their normal ranges.
  - **Changes in Timing and Intensity of Flows:** Climate change can alter the timing and intensity of river flows, which are critical for fish migration, spawning, and access to food resources.
- **Explanation:** Changes in precipitation patterns, a key consequence of climate change, have a direct impact on the water levels and flow rates of inland water bodies. For instance, a river that historically had predictable seasonal floods may now experience more erratic flow patterns, disrupting the life cycles of the fish that depend on that predictability.

13.3.3 Habitat Loss and Degradation

- **Impacts**
  - **Loss of Wetlands:** Wetlands are crucial habitats for many fish species, serving as spawning grounds, nurseries, and feeding areas. Changes in water levels and increased frequency of extreme weather events can degrade or destroy these habitats.

    - **Changes in Lake Levels:** Fluctuations in lake levels due to altered precipitation and evaporation rates can affect spawning sites and shoreline vegetation, which provide important cover and food sources for fish.
    - **Increased Sedimentation and Turbidity:** Increased erosion from more intense rainfall can lead to higher levels of sedimentation and turbidity in rivers and lakes, which can harm fish by clogging their gills, reducing light penetration, and smothering spawning areas.
- **Explanation:** Climate change can alter the physical structure and quality of freshwater habitats. For example, a prolonged drought might cause a lake to shrink, reducing the area of suitable habitat for fish and making them more vulnerable to predators.

13.3.4 Changes in Water Quality

- **Impacts**
    - **Reduced Dissolved Oxygen:** Warmer water holds less dissolved oxygen, which is essential for fish survival. Increased stratification of lakes (where water layers don't mix) can also lead to oxygen depletion in deeper waters.
    - **Increased Pollution:** Changes in rainfall patterns can increase the runoff of pollutants, such as nutrients, pesticides, and heavy metals, into freshwater systems. These pollutants can harm fish directly or indirectly by causing algal blooms and other water quality problems.
    - **Harmful Algal Blooms:** Warmer water temperatures and increased nutrient runoff can create ideal conditions for harmful algal blooms. These blooms can produce toxins that kill fish and other aquatic organisms, and they can also deplete oxygen levels in the water.
- **Explanation:** Climate change can exacerbate existing water quality problems or create new ones. For instance, increased flooding can wash large amounts of nutrients from agricultural areas into rivers, leading to eutrophication and oxygen depletion, making the water uninhabitable for many fish species.

13.3.5 Altered Food Webs

- **Impacts**
    - **Changes in Primary Productivity:** Climate change can affect the growth and abundance of phytoplankton and aquatic plants, which form the base of the food web in many freshwater ecosystems. Changes in temperature, light availability, and nutrient levels can disrupt the entire food web.

  - **Mismatches in Timing:** Changes in water temperature and flow can alter the timing of biological events, such as the emergence of insects or the spawning of fish. These changes can create mismatches between the availability of food and the needs of fish, particularly during critical early life stages.
  - **Invasions of Non-Native Species:** Climate change can make some freshwater ecosystems more hospitable to non-native species, which can outcompete native fish for resources, prey on them, or introduce new diseases.
- **Explanation:** Climate change doesn't just affect fish directly; it also affects the complex web of interactions between fish and other organisms in their environment. For example, if a fish species' primary food source declines due to climate change, that fish population may also decline, even if the fish is otherwise able to tolerate the changing conditions.

13.3.6 Impacts on Fish Migration

- **Impacts**
  - **Disrupted migration patterns:** Changes in water temperature and flow can disrupt the timing and success of fish migration, particularly for species that migrate to spawn.
  - **Barriers to migration:** Extreme weather events and habitat degradation can create physical barriers that prevent fish from reaching their spawning grounds.
- **Explanation:** Many inland fish species undertake migrations within rivers or between rivers and lakes to complete their life cycle. Climate change can interfere with these migrations, for example, by causing a river to dry up during a critical migration period, or by altering the temperature cues that trigger migration.

13.4 Effect of Climate Change on Aquaculture

Inland fisheries, which occur in freshwater environments like rivers, lakes, and wetlands, are vital sources of food and livelihoods for millions of people worldwide. However, these fisheries are particularly vulnerable to the impacts of climate change (Fig. 13.2). Here's a detailed explanation of the effects:

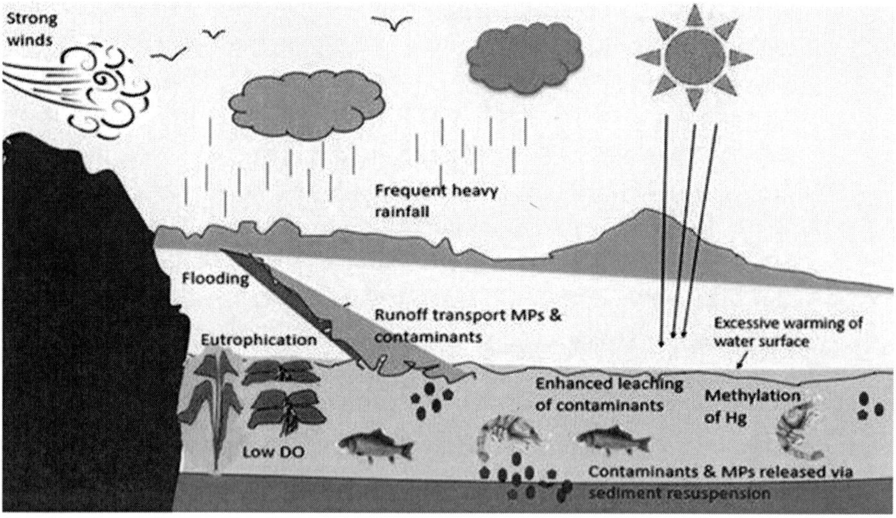

Fig. 13.2: Climate change impact on aquaculture

13.4.1 Changes in Water Temperature

- **Impacts**
  - **Direct Physiological Effects:** Fish are cold-blooded, so their body temperature and metabolic rate are directly influenced by water temperature. Rising temperatures can increase their energy needs, alter growth rates, and affect reproductive success.
  - **Habitat Suitability:** Many fish species have specific temperature requirements. Warming waters can force them to migrate to find suitable habitats, potentially disrupting ecosystems and fisheries. Some species may be unable to relocate and face local extinction.
  - **Increased Stress and Disease:** Higher temperatures can increase stress on fish, making them more susceptible to diseases and parasites.
- **Explanation:** As global air temperatures rise, so do the temperatures of freshwater bodies. This seemingly small change can have drastic effects on the physiology and behavior of fish. For example, in some temperate regions, warmer winters may lead to earlier spawning, which could be mismatched with food availability for newly hatched fish.

13.4.2 Altered Water Flow Regimes

- **Impacts**
  - **Droughts:** Increased evaporation and altered rainfall patterns can lead to more frequent and severe droughts, reducing water levels in

rivers and lakes. This can destroy fish habitats, concentrate pollutants, and limit fish movement.

- **Floods:** Conversely, more intense rainfall events can cause more frequent and severe floods. These can damage fish habitats, disrupt spawning, and wash fish away from their normal ranges.
- **Changes in Timing and Intensity of Flows:** Climate change can alter the timing and intensity of river flows, which are critical for fish migration, spawning, and access to food resources.

- **Explanation:** Changes in precipitation patterns, a key consequence of climate change, have a direct impact on the water levels and flow rates of inland water bodies. For instance, a river that historically had predictable seasonal floods may now experience more erratic flow patterns, disrupting the life cycles of the fish that depend on that predictability.

13.4.3 Habitat Loss and Degradation

- **Impacts**
  - **Loss of Wetlands:** Wetlands are crucial habitats for many fish species, serving as spawning grounds, nurseries, and feeding areas. Changes in water levels and increased frequency of extreme weather events can degrade or destroy these habitats.
  - **Changes in Lake Levels:** Fluctuations in lake levels due to altered precipitation and evaporation rates can affect spawning sites and shoreline vegetation, which provide important cover and food sources for fish.
  - **Increased Sedimentation and Turbidity:** Increased erosion from more intense rainfall can lead to higher levels of sedimentation and turbidity in rivers and lakes, which can harm fish by clogging their gills, reducing light penetration, and smothering spawning areas.
- **Explanation:** Climate change can alter the physical structure and quality of freshwater habitats. For example, a prolonged drought might cause a lake to shrink, reducing the area of suitable habitat for fish and making them more vulnerable to predators.

13.4.4 Changes in Water Quality

- **Impacts**
  - **Reduced Dissolved Oxygen:** Warmer water holds less dissolved oxygen, which is essential for fish survival. Increased stratification

of lakes (where water layers don't mix) can also lead to oxygen depletion in deeper waters.

  - **Increased Pollution:** Changes in rainfall patterns can increase the runoff of pollutants, such as nutrients, pesticides, and heavy metals, into freshwater systems. These pollutants can harm fish directly or indirectly by causing algal blooms and other water quality problems.
  - **Harmful Algal Blooms:** Warmer water temperatures and increased nutrient runoff can create ideal conditions for harmful algal blooms. These blooms can produce toxins that kill fish and other aquatic organisms, and they can also deplete oxygen levels in the water.
- **Explanation:** Climate change can exacerbate existing water quality problems or create new ones. For instance, increased flooding can wash large amounts of nutrients from agricultural areas into rivers, leading to eutrophication and oxygen depletion, making the water uninhabitable for many fish species.

13.4.5 Altered Food Webs

- **Impacts**
  - **Changes in Primary Productivity:** Climate change can affect the growth and abundance of phytoplankton and aquatic plants, which form the base of the food web in many freshwater ecosystems. Changes in temperature, light availability, and nutrient levels can disrupt the entire food web.
  - **Mismatches in Timing:** Changes in water temperature and flow can alter the timing of biological events, such as the emergence of insects or the spawning of fish. These changes can create mismatches between the availability of food and the needs of fish, particularly during critical early life stages.
  - **Invasions of Non-Native Species:** Climate change can make some freshwater ecosystems more hospitable to non-native species, which can outcompete native fish for resources, prey on them, or introduce new diseases.
- **Explanation:** Climate change doesn't just affect fish directly; it also affects the complex web of interactions between fish and other organisms in their environment. For example, if a fish species' primary food source declines due to climate change, that fish population may also decline, even if the fish is otherwise able to tolerate the changing conditions.

13.4.6 Impacts on Fish Migration

- Impacts
  - **Disrupted migration patterns:** Changes in water temperature and flow can disrupt the timing and success of fish migration, particularly for species that migrate to spawn.
  - **Barriers to migration:** Extreme weather events and habitat degradation can create physical barriers that prevent fish from reaching their spawning grounds.
- **Explanation:** Many inland fish species undertake migrations within rivers or between rivers and lakes to complete their life cycle. Climate change can interfere with these migrations, for example, by causing a river to dry up during a critical migration period, or by altering the temperature cues that trigger migration.

13.5 Soil and Water Pollution Effects on Fisheries

Soil and water pollution have profound and detrimental effects on fisheries, impacting aquatic ecosystems and the health and productivity of fish populations. Here's a breakdown of the key ways pollution affects fisheries:

13.5.1 Types of Pollution and Their Sources

- **Sediment Pollution**
  - **Sources:** Soil erosion from deforestation, agriculture, construction, and mining activities.
  - **Impacts:** Increased turbidity, which reduces light penetration, hindering photosynthesis and disrupting the food chain. Sediment can also smother spawning grounds, clog fish gills, and transport other pollutants.
- **Nutrient Pollution**
  - **Sources:** Agricultural runoff (fertilizers), sewage discharge.
  - **Impacts:** Eutrophication, leading to excessive algae growth. Algal blooms deplete dissolved oxygen when they decompose, creating "dead zones" where fish and other aquatic life cannot survive.
- **Toxic Chemical Pollution**
  - **Sources:** Industrial discharge, pesticides, herbicides, heavy metals, oil spills.

- **Impacts:** Direct toxicity to fish, causing mortality, reproductive impairment, and disease. Bioaccumulation and biomagnification of toxins in the food chain can make fish unsafe for consumption.

- **Acidification**
  - **Sources:** Acid rain from atmospheric pollution (sulfur dioxide and nitrogen oxides), acid mine drainage.
  - **Impacts:** Lower pH in water bodies, harming fish directly (especially eggs and larvae) and disrupting their physiological processes. Acidification also releases toxic metals from sediments.

- **Organic Pollution**
  - **Sources:** Sewage, livestock waste, food processing waste.
  - **Impacts:** Depletion of dissolved oxygen as organic matter decomposes, harming fish and other aquatic organisms. Pathogens in organic waste can also cause disease in fish and humans.

- **Thermal Pollution**
  - **Sources:** Discharge of heated water from industrial processes, power plants.
  - **Impacts:** Altered water temperatures, which can disrupt fish metabolism, reproduction, and migration patterns. Thermal pollution can also decrease dissolved oxygen levels.

13.5.2 Impacts on Fish and Aquatic Ecosystems

- **Direct Toxicity:** Many pollutants are directly toxic to fish, causing a range of effects:
  - **Mortality:** High concentrations of pollutants can lead to large-scale fish kills.
  - **Physiological Damage:** Pollutants can damage fish gills, liver, kidneys, and nervous systems.
  - **Reproductive Impairment:** Pollutants can interfere with fish reproduction, reducing fertility, hatching success, and larval survival.
  - **Disease:** Pollution can weaken fish immune systems, making them more susceptible to diseases.
- **Habitat Destruction:** Pollution can degrade or destroy fish habitats:
  - **Smothering of Spawning Grounds:** Sediment pollution can cover eggs and prevent them from hatching.

- **Loss of Vegetation:** Turbidity and toxic chemicals can kill aquatic plants, which provide food and shelter for fish.
- **Alteration of Water Flow:** Pollution that changes sedimentation patterns or increases flooding can destroy habitats.

- **Disruption of Food Webs:** Pollution can disrupt the balance of aquatic ecosystems:
  - **Reduced Primary Productivity:** Turbidity and toxic chemicals can harm phytoplankton and aquatic plants, reducing the base of the food web.
  - **Loss of Biodiversity:** Pollution can eliminate sensitive species, reducing the diversity and stability of the ecosystem.
  - **Bioaccumulation and Biomagnification:** Toxic pollutants can accumulate in the tissues of fish and other aquatic organisms, and their concentrations can increase as they move up the food chain, harming top predators and humans who consume contaminated fish.
- **Oxygen Depletion**
  - **Eutrophication:** Nutrient pollution can cause algal blooms, which deplete oxygen when they decompose, creating hypoxic or anoxic conditions that kill fish.
  - **Organic Matter Decomposition:** The breakdown of organic pollutants also consumes oxygen, further exacerbating oxygen depletion.
- **Increased Susceptibility to Other Stressors:** Polluted environments can make fish more vulnerable to other stressors, such as:
  - **Climate Change:** Polluted fish populations may be less able to tolerate rising temperatures and altered water flows.
  - **Overfishing:** Pollution can reduce fish populations, making them more susceptible to overfishing.

13.5.3 Effects on Fisheries and Human Communities

- **Reduced Fish Stocks:** Pollution can lead to significant declines in fish populations, reducing catches for commercial and recreational fisheries.
- **Economic Losses:** Fisheries closures, reduced catches, and contaminated fish can cause substantial economic losses for fishing communities and related industries.
- **Food Security Impacts:** In communities that rely heavily on fish for protein, pollution can threaten food security and nutrition.

- **Health Risks:** Consumption of contaminated fish can expose humans to toxins, leading to various health problems.
- **Loss of Livelihoods:** Pollution can harm the livelihoods of fishermen, fish farmers, and others who depend on healthy aquatic ecosystems.

13.6 Strategies for Climate Resilience in Fisheries Sector

13.6.1 Climate Resilience in Marine Fisheries

Marine fisheries are particularly vulnerable to climate change due to the impacts of rising sea temperatures, ocean acidification, and sea-level rise. Here are some key strategies:

- **Ecosystem-Based Fisheries Management (EBFM)**
  - Move beyond single-species management to consider the entire ecosystem, including habitats, food webs, and human impacts.
  - Implement spatial management tools like marine protected areas (MPAs) to conserve critical habitats and allow fish populations to recover.
  - Promote sustainable fishing practices that minimize bycatch and habitat damage.
- **Flexible and Adaptive Management**
  - Develop management plans that can be easily adjusted in response to changing climate conditions.
  - Invest in robust monitoring and data collection systems to track changes in fish stocks and ecosystems.
  - Use climate models and forecasting to anticipate future impacts and inform management decisions.
- **Diversification of Fishing Livelihoods**
  - Reduce reliance on a single fishery by promoting alternative livelihoods for fishing communities, such as aquaculture, tourism, or other marine-related activities.
  - Provide training and support for fishers to transition to new livelihoods.
- **Strengthening Coastal Resilience**
  - Protect and restore coastal habitats like mangroves, coral reefs, and seagrass beds, which provide important nursery grounds for fish and buffer coastlines from storms and sea-level rise.

- Develop coastal management plans that account for climate change impacts and prioritize ecosystem-based adaptation measures.

- **Addressing Ocean Acidification**
  - Support research to understand the impacts of ocean acidification on marine organisms and ecosystems.
  - Implement measures to reduce CO_2 emissions, the primary driver of ocean acidification.
  - Explore adaptation strategies, such as selective breeding of more resilient species.

13.6.2 Climate Resilience in Inland Fisheries

Inland fisheries are affected by changes in water temperature, altered hydrological regimes, and habitat degradation. Here are some resilience strategies:

- **Integrated Water Resources Management (IWRM)**
  - Manage water resources in a holistic way, considering the needs of fisheries along with other users like agriculture, industry, and domestic water supply.
  - Ensure adequate water flows in rivers and maintain healthy water levels in lakes and wetlands to support fish populations.
  - Implement measures to improve water quality, such as reducing pollution from agricultural runoff and sewage discharge.
- **Habitat Restoration and Connectivity**
  - Restore degraded freshwater habitats, such as rivers, lakes, and wetlands, to improve their capacity to support fish.
  - Remove barriers to fish migration, such as dams and culverts, to allow fish to access spawning and feeding grounds.
  - Protect and restore riparian vegetation, which helps to stabilize riverbanks, filter pollutants, and provide shade that regulates water temperature.
- **Climate-Smart Fisheries Management**
  - Incorporate climate change projections into fisheries management plans.
  - Establish protected areas to safeguard critical fish habitats and migration routes.

    - Implement flexible fishing regulations that can be adjusted in response to changing environmental conditions.

- **Diversification of Livelihoods**
    - Promote alternative livelihoods for fishing communities, such as aquaculture, eco-tourism, or sustainable agriculture.
- **Flood and Drought Preparedness**
    - Develop early warning systems for floods and droughts to help fishing communities prepare for extreme weather events.
    - Implement flood-resilient infrastructure and practices, such as constructing fish farms in less flood-prone areas and promoting flood-recession agriculture.

13.6.3 Climate Resilience in Aquaculture

Aquaculture faces challenges from changing water temperatures, extreme weather events, and disease outbreaks. Here's how to enhance resilience:

- **Diversification of Species and Systems**
    - Culture a variety of species with different climate tolerances to reduce vulnerability to specific climate impacts.
    - Adopt diverse aquaculture systems, such as integrated multi-trophic aquaculture (IMTA), which can improve water quality and reduce waste.
- **Climate-Smart Farm Management**
    - Select farm sites that are less vulnerable to climate change impacts, such as flooding, sea-level rise, and extreme temperatures.
    - Improve water management practices to conserve water and mitigate the impacts of droughts and floods.
    - Implement biosecurity measures to prevent and manage disease outbreaks, which can be exacerbated by climate change.
- **Technology and Innovation**
    - Invest in research and development to improve the climate resilience of aquaculture species and systems.
    - Develop and adopt new technologies, such as climate-controlled aquaculture systems and early warning systems for extreme weather events.

- **Policy and Governance**
  - Develop policies and regulations that promote climate-resilient aquaculture practices.
  - Incorporate aquaculture into national climate change adaptation plans.
  - Provide financial incentives and support for farmers to adopt climate-resilient practices.
- **Insurance and Risk Management**
  - Develop insurance schemes to help aquaculture farmers cope with losses from climate-related events.
  - Promote risk management strategies, such as diversifying farm operations and developing contingency plans.

Effect of Climate Change on the Fisheries Sector

1. Which greenhouse gas is most closely linked with global warming?

a) Nitrogen
b) Oxygen
c) Carbon dioxide
d) Argon

2. Which sector is directly affected by ocean acidification?

a) Mining
b) Transport
c) Fisheries
d) Construction

3. Which phenomenon causes bleaching of corals due to rising sea temperatures?

a) Eutrophication
b) Coral respiration
c) Coral bleaching
d) Sediment loading

4. Ocean acidification is mainly due to:

a) Carbon monoxide
b) CO_2 absorption by seawater
c) Oil spills
d) Decreased rainfall

5. Which climatic factor most influences fish migration?

a) Rainfall
b) Temperature
c) Wind speed
d) Fog

6. The El Niño phenomenon results in:

a) Cooling of ocean waters
b) Stabilization of salinity
c) Warming of surface waters in the Pacific
d) Increased snow cover

7. Climate change is expected to reduce fish availability due to:

a) Better gear
b) Nutrient enrichment
c) Habitat destruction and migration shifts
d) Increased fishermen activity

8. Which group of fish is highly sensitive to temperature changes?

a) Mollusks
b) Pelagic fish
c) Cartilaginous fish
d) Benthos

9. Sea level rise can result in:

a) New estuaries
b) Loss of coastal wetlands
c) Decrease in salinity
d) Reduced wave action

10. Increased storm intensity due to climate change threatens fisheries by:

a) Enhancing oxygen
b) Damaging gear and infrastructure
c) Improving catch potential
d) Increasing fish quality

11. Warming oceans affect fish by altering their:

a) Color
b) Taste
c) Distribution and breeding grounds
d) Bone structure

12. Ocean deoxygenation refers to:

a) Water purification
b) Loss of oxygen in deeper waters
c) Photosynthesis in fish
d) CO_2 exhalation by coral

13. Lower dissolved oxygen in warmer water leads to:

a) Faster growth
b) Improved health
c) Fish mortality and stress
d) Higher profits

14. Which species is more vulnerable to ocean acidification?

a) Sharks
b) Crustaceans
c) Bony fish
d) Dolphins

15. Shifting fish stocks may lead to:

a) Stability in income
b) Abundance in all regions
c) Conflict among fishing communities
d) Increase in mangroves

16. Coral reef destruction impacts fisheries because:

a) Reefs clean water
b) Reefs are breeding habitats for many fish
c) Reefs reduce waves
d) Reefs produce oxygen

17. Which indicator shows ocean warming due to climate change?

a) Increase in sedimentation
b) Decrease in sea surface temperature
c) Coral bleaching events
d) Reduced salinity

18. Temperature fluctuations in hatcheries can:

a) Improve survival
b) Slow bacterial growth
c) Cause deformities and low yield
d) Increase nutrient content

19. Which oceanic change due to climate affects phytoplankton growth?

a) Decreased evaporation
b) Reduced light penetration
c) Nutrient cycling due to stratification
d) Magnetic flux

20. Fish species moving toward poles is a response to:

a) Higher pressure
b) Lower oxygen
c) Warming temperatures
d) Increasing plankton

21. Why are estuarine fish species vulnerable to climate change?

a) They are migratory
b) They depend on narrow salinity ranges
c) They are herbivores
d) They reproduce rapidly

22. Shellfish industry is affected most by:

a) Algal blooms
b) Ocean acidification
c) Wind shear
d) Stratification

23. Extreme rainfall affects inland aquaculture by:

a) Enhancing feed quality
b) Causing pond flooding and fish escapes
c) Reducing oxygen
d) Increasing alkalinity

24. The migration patterns of fish are influenced by:

a) Tidal cycles
b) Temperature and ocean currents
c) Seabed hardness
d) Moonlight

25. Climate-induced diseases in aquaculture are caused by:

a) Noise
b) Thermal stress and water quality changes
c) Fish overcrowding only
d) Declining feed supply

26. Which practice supports adaptation to climate change in aquaculture?

a) Stocking wild fish only
b) Monoculture systems
c) Polyculture and species diversification
d) Complete reliance on rain

27. Resilient aquaculture systems are:

a) Less profitable
b) Highly dependent on fertilizers
c) Capable of withstanding climate shocks
d) Always offshore

28. Mangroves help fisheries by:

a) Polluting the water
b) Absorbing CO_2 and providing fish nursery grounds
c) Reducing light
d) Increasing soil erosion

29. Climate-smart aquaculture involves:

a) Overuse of antibiotics
b) Higher chemical usage
c) Efficient water and energy use with adaptive practices
d) Ignoring temperature changes

30. Mitigation in fisheries focuses on:

a) Reducing methane
b) Fishing harder
c) Lowering carbon emissions from fishing fleets
d) Burning fuel

31. Early warning systems for fishers aim to:

a) Enhance breeding
b) Improve safety and planning
c) Predict oil prices
d) Locate fishing licenses

32. What is carbon sequestration in aquatic systems?

a) Emission of smoke
b) Removal of nitrogen
c) Capturing carbon in plants like mangroves
d) Increasing methane

33. Which farming system helps reduce climate vulnerability?

a) Extensive cage culture in oceans
b) Integrated fish-livestock-farming systems
c) Exclusive monoculture
d) Floating plastic tanks

34. Which is a climate change adaptation technology in aquaculture?

a) Solar-powered aerators
b) Fossil fuel engines
c) Deep water drilling
d) Fish importation

35. Relocation of hatcheries inland is an adaptation to:

a) Heat stress
b) Urbanization
c) Sea level rise
d) Light pollution

36. Selective breeding in fish helps in:

a) Faster marketing
b) Enhancing resistance to environmental stress
c) Lowering plankton levels
d) Removing parasites

37. Diversification of income for fishers includes:

a) Online gaming
b) Tourism, crafts, aquaculture
c) Agriculture only
d) Boat trading

38. International agreements to tackle climate change include:

a) Geneva Protocol
b) Paris Agreement
c) FAO Code of Conduct
d) Kyoto Treaty only

39. The vulnerability of a fishery sector depends on:

a) Political leadership only
b) Climate exposure, sensitivity, and adaptive capacity
c) Cost of nets
d) Depth of water

40. Which species can be climate-resilient in aquaculture?

a) Those with narrow thermal tolerance
b) Coldwater species
c) Air-breathing and hardy species
d) Deep-sea fish only

41. Which adaptive strategy helps reduce the vulnerability of fishers to climate change?

a) Use of outdated fishing gear
b) Diversification of income sources
c) Monoculture of aquatic species
d) Overfishing during monsoon

42. Which farming system promotes resilience in aquaculture against climate change?

a) Single-species culture
b) Integrated fish farming systems
c) Pure crop farming
d) Industrial fishing

43. What role do mangroves play in climate adaptation for fisheries?

a) Increase water temperature
b) Provide nursery grounds and buffer storm surges
c) Encourage pollution
d) Decrease salinity

44. Which of the following is a major research focus in climate-resilient aquaculture?

a) Enhancing boat speed
b) Developing heat-tolerant and disease-resistant fish strains
c) Promoting plastic cage farming
d) Artificial coloring of fish

45. The role of meteorological services in fisheries adaptation is to:

a) Predict fish prices
b) Offer investment advice
c) Provide early warnings and weather forecasts
d) Delay harvesting decisions

46. One of the SDG (Sustainable Development Goals) most closely linked to climate and fisheries is:

a) SDG 1: No Poverty
b) SDG 14: Life Below Water
c) SDG 11: Sustainable Cities
d) SDG 9: Industry and Infrastructure

47. Ocean acidification caused by climate change directly affects:

a) Fishing licenses
b) Shell-forming organisms and coral reefs
c) Boat colors
d) Plastic nets

48. What is a co-benefit of adopting climate-smart aquaculture practices?

a) Higher market tax
b) Increased disease outbreaks
c) Resource efficiency and emission reduction
d) Overstocking ponds

49. Which international organization supports climate adaptation in fisheries?

a) WHO
b) FAO
c) WTO
d) UNESCO

50. The ultimate goal of climate adaptation strategies in fisheries is to:

a) Reduce market price of fish
b) Maximize short-term yield
c) Ensure long-term sustainability and livelihoods
d) Increase seafood imports

Answer Key (s)

1	c	2	c	3	c	4	b	5	b	6	c	7	c	8	b	9	b	10	b
11	c	12	b	13	c	14	b	15	c	16	b	17	c	18	c	19	c	20	c
21	b	22	b	23	b	24	b	25	b	26	c	27	c	28	b	29	c	30	c
31	b	32	c	33	b	34	a	35	c	36	b	37	b	38	b	39	b	40	c
41	b	42	b	43	b	44	b	45	c	46	b	47	b	48	c	49	b	50	c

14

Useful Instruments

Some of the instruments commonly used for taking/recording weather observations are described below:

14.1. Sixs Maximum and Minimum Thermometer

The **Six's maximum and minimum thermometer** is a U-shaped glass tube used to record the highest and lowest temperatures reached over a period, typically 24 hours. It was invented by James Six in the late 18th century, and its basic design is still in use today (Fig. 14.1).

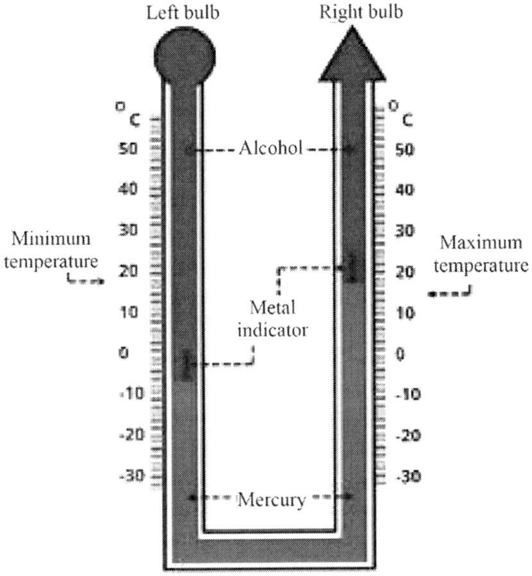

Fig. 14.1: Six's maximum and minimum thermometer

14.1.1 Description

- It consists of a U-shaped glass tube with a bulb at each end.
- One bulb is partially filled with alcohol (or sometimes creosote), and the space above it is a vacuum.

- The bend of the U-tube contains a thread of mercury.
- In each arm of the U-tube, above the mercury, there is a small, movable index (often made of steel) with a small spring that allows it to be pushed by the mercury but remain in its position when the mercury retreats.
- Along each arm of the U-tube, there is a temperature scale. The scale on the maximum temperature side increases upwards, while the scale on the minimum temperature side increases downwards (inverted).

14.1.2 How it Works

- **Temperature Rise:** When the temperature rises, the alcohol in the bulb expands, pushing the mercury up the arm on the maximum temperature side. The mercury pushes the index upwards, and this index stays at the highest point reached as the mercury falls back down with decreasing temperature.
- **Temperature Fall:** When the temperature falls, the alcohol in the bulb contracts, pulling the mercury down on the minimum temperature side (and thus up on the maximum side). The mercury on the minimum side pushes its index upwards. As the temperature rises again, the mercury retreats, leaving the minimum temperature index at the lowest point reached.
- **Reading the Thermometer:** To read the maximum temperature, you look at the lower end of the index on the maximum temperature scale. To read the minimum temperature, you look at the lower end of the index on the minimum temperature scale. The current temperature can be read from either side by noting the position of the top of the mercury column.
- **Resetting the Thermometer:** After taking the readings, the indices need to be reset. This is typically done using a small magnet to pull the steel indices back down to the current level of the mercury.

14.2 Assmans Psychrometer

14.2.1 Construction of the Assnann's Psychrometer

The Assmann psychrometer is a precision instrument designed for accurate measurement of air humidity (Fig. 14.2). Its key features include:

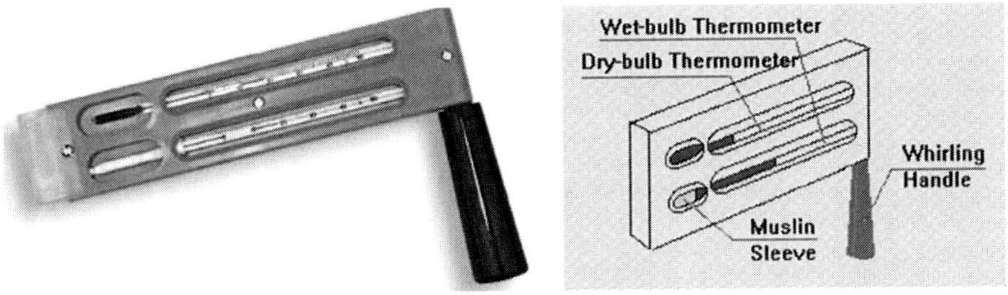

Fig. 14.2: Assmann psychrometer

- **Two Thermometers:** It utilizes two matched mercury-in-glass thermometers mounted side-by-side.
  - **Dry-bulb thermometer:** Measures the ambient air temperature. Its bulb is exposed directly to the air.
  - **Wet-bulb thermometer:** Measures the temperature of evaporation. Its bulb is covered with a clean muslin wick that is moistened with distilled water before use.
- **Radiation Shields:** Each thermometer bulb is encased in a double-walled, polished metallic tube (radiation shield). This design minimizes the influence of direct solar radiation and long-wave radiation from surrounding objects, ensuring that the thermometers primarily respond to the air temperature and the effect of evaporation.
- **Aspirator (Fan):** A spring-driven or battery-powered fan is integrated into the instrument. This fan draws air at a constant and known velocity (typically around 2-3 m/s) over the bulbs of both thermometers. This forced ventilation ensures maximum evaporation from the wet bulb and provides a representative sample of the surrounding air to both thermometers, leading to more accurate and consistent readings compared to naturally ventilated psychrometers.
- **Protective Casing:** The thermometers and the aspirator mechanism are housed within a sturdy casing, often metallic, which also serves as a handle for holding or mounting the instrument.
- **Moistening Device:** A small container or syringe is usually provided for wetting the muslin wick of the wet-bulb thermometer with distilled water.

14.2.2 Working and Use for Measuring Relative Humidity

The Assmann psychrometer works on the principle of evaporative cooling. Here's how it's used to determine relative humidity:

- **Wetting the Wick:** The muslin wick covering the bulb of the wet-bulb thermometer is thoroughly moistened with distilled water. Excess water should be avoided.
- **Activating the Aspirator:** The fan (aspirator) is started (either by winding a spring or turning it on if battery-powered) to draw air over the thermometer bulbs at a constant rate.
- **Taking Readings:** After a few minutes of ventilation, the temperature of the wet-bulb thermometer will stabilize due to the cooling effect of evaporation. Simultaneously, the dry-bulb thermometer will register the ambient air temperature. The readings of both thermometers are carefully recorded.
- **Determining Relative Humidity:** The relative humidity is then determined using the readings from the dry-bulb temperature (Td) and the wet-bulb temperature (Tw). This is typically done in one of two ways:
  - **Psychrometric Tables:** Specially prepared tables relate the dry-bulb temperature and the difference between the dry-bulb and wet-bulb temperatures (Td −Tw, known as the psychrometric depression) to the relative humidity. You would look up the Td and the calculated depression to find the corresponding relative humidity percentage.
  - **Psychrometric Formulas:** Empirical formulas, based on the physics of evaporation and the properties of air and water vapor, can be used to calculate the vapor pressure and subsequently the relative humidity using Td , Tw , and the atmospheric pressure.

14.2.3 Why the Assmann Psychrometer is Accurate

- **Forced Ventilation:** The built-in fan ensures consistent and maximum evaporation, leading to a well-defined wet-bulb temperature.
- **Radiation Shielding:** The double-walled radiation shields minimize errors caused by radiant heat exchange.8
- **Matched Thermometers:** Using calibrated, matched thermometers improves the accuracy of the temperature difference measurement.

In summary, the Assmann psychrometer is a reliable instrument for measuring relative humidity by accurately determining the dry-bulb and wet-bulb temperatures through forced ventilation and radiation shielding, and then using these values with psychrometric tables or formulas.

14.3 Aneroid Barometer

14.3.1 Construction

- An aneroid barometer uses a small, flexible metal box called an **aneroid cell** (or vacuum chamber). This box is partially evacuated, meaning it contains very little air (Fig. 14.3).
- The sides of the aneroid cell are sensitive to changes in external pressure.
- A system of levers and springs is connected to the aneroid cell. As the atmospheric pressure changes, the cell either expands or contracts. These movements are amplified by the levers and transmitted to a pointer that moves across a calibrated dial.
- Some modern aneroid barometers use a transducer to convert the mechanical movement into an electronic signal for digital displays or automated data logging.

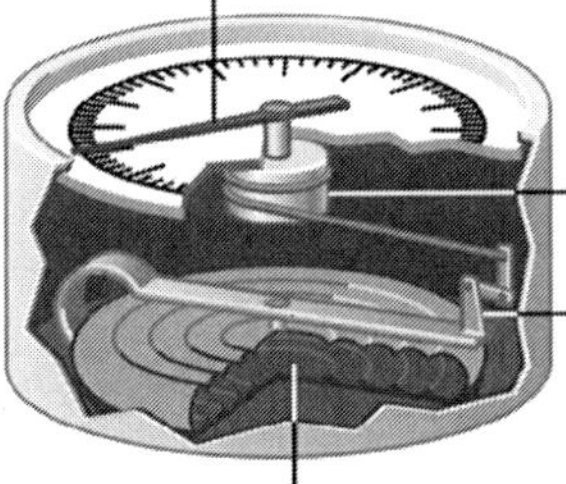

Fig. 14.3: Aneroid barometer

14.3.2 Working Principle

- When the atmospheric pressure increases, it presses on the outside of the aneroid cell, causing it to compress slightly.
- When the atmospheric pressure decreases, the internal spring of the aneroid cell pulls the sides outward, causing it to expand.
- These changes in the shape of the aneroid cell are mechanically linked to the pointer on the dial, indicating the corresponding pressure.

14.3.3 Measurement

- The atmospheric pressure is directly read from the calibrated dial in units such as hectopascals (hPa), millibars (mb), or inches of mercury (inHg).

- Aneroid barometers are more portable and robust than mercury barometers but may require periodic calibration against a more accurate standard.

14.4 Thermograph

Let's discuss the construction, working principle, and maintenance of a **thermograph**, which is an instrument that continuously records temperature over time, typically on a chart.

14.4.1 Construction: A Mechanical Thermograph Typically Consists of the Following Key Components

- **Temperature-Sensing Element:** The most common sensing element is a **bimetallic strip** or a **bimetallic helix**. This element is made of two different metals with different coefficients of thermal expansion (how much they expand or contract with temperature changes) bonded together.1 As the temperature changes, the differential expansion of the two metals causes the strip or helix to bend or coil.
- **Amplifying Linkage:** The movement of the bimetallic strip or helix is often small. A system of mechanical linkages (levers) is used to amplify this movement.
- **Recording Arm and Pen:** The amplified movement is transferred to a recording arm that holds a pen (often a fiber-tipped pen or a stylus).
- **Rotating Drum and Chart:** A cylindrical drum, driven by a clockwork or electric motor, slowly rotates. A chart, graduated with temperature on the vertical axis and time on the horizontal axis, is wrapped around this drum.
- **Protective Casing:** All these components are usually housed within a protective casing with ventilation to allow ambient air to reach the sensing element while shielding it from direct sunlight and precipitation.

14.4.2 Working Principle

The working principle of a mechanical thermograph relies on the **differential thermal expansion** of the bimetallic strip or helix:

- **Temperature Change Leads to Movement:** When the ambient temperature changes, the two metals in the bimetallic strip expand or contract by different amounts.
- **Bending or Coiling:** This differential expansion causes the bimetallic strip to bend or the helix to coil or uncoil. The amount of bending or coiling is proportional to the change in temperature.

- **Amplification and Recording:** The mechanical linkage amplifies this movement and transmits it to the recording arm holding the pen. As the drum rotates at a constant speed, the pen moves up or down on the chart in response to temperature changes, tracing a continuous record of temperature over time. The resulting line on the chart is called a **thermogram**.

14.4.3 Measurement

The thermogram on the chart provides a visual record of temperature fluctuations. To obtain a specific temperature reading at a particular time:

- Locate the point on the traced line corresponding to the desired time (using the time scale on the horizontal axis).
- Read the corresponding temperature value from the temperature scale on the vertical axis at that point.

The thermogram also allows for easy identification of maximum and minimum temperatures within the recording period and the rate at which temperature changes.

14.4.4 Maintenance of a Thermograph

Regular maintenance is essential for ensuring the accuracy and longevity of a thermograph.Common maintenance tasks include:

- **Chart Replacement:** Regularly replacing the chart (daily, weekly, or monthly, depending on the model) is necessary to have a fresh recording surface. Ensure the new chart is fitted correctly on the drum and aligned with the time markings.
- **Pen Care**
  - **Ink Level:** If using an inked pen, check the ink level regularly and refill or replace the pen as needed to ensure a continuous trace.
  - **Pen Pressure:** Adjust the pen pressure on the chart so that it makes a clear mark without scratching or tearing the paper.
  - **Cleaning:** Clean the pen tip periodically to remove any dried ink or debris that could affect the trace.
- **Clock/Motor Maintenance**
  - For clockwork mechanisms, ensure it is wound regularly.
  - For electric motors, check the power supply and ensure smooth operation of the drum rotation.

- **Calibration:** Periodically, the thermograph should be calibrated against a more accurate standard thermometer (like a calibrated liquid-in-glass thermometer or a digital thermometer). This involves:
    - Placing the sensing element of the thermograph and the standard thermometer in close proximity where the temperature is relatively stable.
    - Comparing the temperature reading on the standard thermometer with the trace on the thermograph at that time.
    - If there is a significant discrepancy, the recording arm or the linkage mechanism might need adjustment according to the manufacturer's instructions to bring the thermogram in line with the standard.
- **Cleaning the Instrument:** Gently clean the exterior of the thermograph to remove dust and dirt. Ensure the ventilation openings are not blocked.
- **Checking the Bimetallic Element:** While less frequent, the bimetallic strip or helix should be visually inspected for any damage or corrosion. However, this is usually a more involved process and might require professional servicing.

By performing these maintenance tasks regularly, you can ensure that the thermograph provides a reliable and continuous record of temperature.

14.5 Barograph

Let's discuss the construction, working principle, and maintenance of a **barograph**, which is an instrument that continuously records atmospheric pressure over time on a chart.

14.5.1 Construction

A mechanical barograph typically consists of the following key components:

- **Pressure-Sensing Element:** The primary sensing element is usually a stack of several **aneroid cells**(also known as vacuum chambers).1 These are partially evacuated, thin metal capsules that are sensitive to changes in atmospheric pressure.2 Using a stack increases the overall movement in response to pressure variations.
- **Amplifying Linkage:** The expansion and contraction of the aneroid cells are often small. A system of mechanical levers is used to amplify these movements.
- **Recording Arm and Pen:** The amplified movement is transferred to a recording arm that holds a pen (often a fiber-tipped pen or a stylus).

- **Rotating Drum and Chart:** Similar to a thermograph, a cylindrical drum, driven by a clockwork or electric motor, slowly rotates. A chart, graduated with pressure on the vertical axis and time on the horizontal axis, is wrapped around this drum.
- **Protective Casing:** The internal components are housed in a protective casing with a mechanism to allow ambient atmospheric pressure to act on the aneroid cells.

14.5.2 Working Principle

The working principle of a mechanical barograph relies on the elastic properties of the aneroid cells responding to changes in atmospheric pressure:

- **Pressure Change Leads to Movement:** When the atmospheric pressure increases, it compresses the aneroid cells, causing them to contract.3 Conversely, when the pressure decreases, the internal springs of the cells cause them to expand.
- **Amplification and Recording:** The mechanical linkage amplifies these small changes in the height of the aneroid cell stack and transmits this amplified movement to the recording arm holding the pen.5 As the drum rotates at a constant speed, the pen moves up or down on the chart in response to pressure changes, tracing a continuous record of atmospheric pressure over time.6 The resulting line on the chart is called a **barogram**.

14.5.3 Measurement

The barogram on the chart provides a visual record of pressure fluctuations.To obtain a specific pressure reading at a particular time:

- Locate the point on the traced line corresponding to the desired time (using the time scale on the horizontal axis).
- Read the corresponding pressure value from the pressure scale on the vertical axis at that point.

The barogram is useful for observing trends in atmospheric pressure, such as the approach of a weather system (often indicated by a falling pressure).

14.5.4 Maintenance of a Barograph

Regular maintenance is important for the accurate and reliable operation of a barograph. Common maintenance tasks include.

- **Chart Replacement:** Regularly replace the chart (daily, weekly, or monthly) to provide a fresh recording surface. Ensure it's fitted correctly and aligned with the time markings.

- **Pen Care**
  - **Ink Level:** If using an inked pen, check and refill or replace it as needed.
  - **Pen Pressure:** Adjust the pen pressure to ensure a clear trace without damaging the chart.
  - **Cleaning:** Clean the pen tip periodically.
- **Clock/Motor Maintenance**
  - Wind the clockwork mechanism regularly.
  - Check the power supply and operation of electric motors.
- **Calibration:** Periodically, the barograph should be calibrated against a more accurate barometer (like a calibrated aneroid or mercury barometer). This involves:
  - Comparing the pressure reading from the standard barometer with the reading indicated by the pen on the barograph chart at that time.
  - If there's a consistent difference, the linkage mechanism of the barograph might need adjustment according to the manufacturer's instructions to align the barogram with the standard. This often involves adjusting a screw or lever that affects the pen's vertical position for a given pressure.
- **Cleaning the Instrument:** Gently clean the exterior of the barograph. Ensure that the mechanism allowing atmospheric pressure to reach the aneroid cells is not obstructed.

By performing these maintenance tasks, you can help ensure that the barograph provides a continuous and reasonably accurate record of atmospheric pressure changes.

14.6 Hair Hygrograph

Let's discuss the construction, working principle, and maintenance of a **hair hygrograph**, which is an instrument that continuously records relative humidity over time on a chart, utilizing the property of hair to change length with humidity.

14.6.1 Construction

A mechanical hair hygrograph typically consists of the following key components:

- **Humidity-Sensing Element:** The primary sensing element is a bundle of **human or animal hairs** (historically, often blonde human hair because it was found to be most responsive). These hairs are hygroscopic, meaning they absorb or lose moisture from the air, causing their length to change. Multiple hairs are used in parallel to increase the force exerted by their expansion and contraction.

- **Tensioning Mechanism:** The hair bundle is kept under a slight tension, usually by a spring. This ensures that any change in length due to humidity variations results in a corresponding movement.
- **Amplifying Linkage:** The change in length of the hair bundle is typically small (a few percent between very dry and very humid conditions). A system of mechanical levers amplifies this change in length.
- **Recording Arm and Pen:** The amplified movement is transferred to a recording arm that holds a pen (often a fiber-tipped pen or a stylus).
- **Rotating Drum and Chart:** Similar to a thermograph and barograph, a cylindrical drum, driven by a clockwork or electric motor, slowly rotates. A chart, graduated with relative humidity (usually in percentage) on the vertical axis and time on the horizontal axis, is wrapped around this drum.
- **Protective Casing:** The internal components are housed within a protective casing with ventilation to allow the ambient air to circulate around the hair bundle.

14.6.2 Working Principle

The working principle of a hair hygrograph relies on the **hygroscopic property of hair**.

- **Humidity Change Leads to Length Change:** As the relative humidity of the surrounding air changes, the hairs absorb or lose moisture.This absorption or loss of moisture causes the length of the hair bundle to change. The length increases with increasing humidity and decreases with decreasing humidity.
- **Amplification and Recording:** The tensioned hair bundle's change in length causes a movement in the mechanical linkage. This movement is amplified and transmitted to the recording arm holding the pen. As the drum rotates, the pen moves up or down on the chart in response to changes in the length of the hair bundle, thus tracing a continuous record of relative humidity over time. The resulting line on the chart is called a **hygrogram**.

14.6.3 Measurement

The hygrogram on the chart provides a visual record of relative humidity fluctuations. To obtain a specific relative humidity reading at a particular time:

- Locate the point on the traced line corresponding to the desired time (using the time scale on the horizontal axis).
- Read the corresponding relative humidity value from the humidity scale (usually in %) on the vertical axis at that point.

The hygrogram shows how relative humidity changes over the recording period, including any peaks and troughs.

14.6.4 Maintenance of a Hair Hygrograph

Regular maintenance is crucial for maintaining the accuracy of a hair hygrograph. Common maintenance tasks include:

- **Chart Replacement:** Regularly replace the chart (daily, weekly, or monthly) and ensure proper fitting and alignment with the time markings.
- **Pen Care:** Maintain the pen's ink level, pressure, and cleanliness for a clear trace.
- **Clock/Motor Maintenance:** Ensure the drum rotation mechanism (clockwork or motor) is functioning correctly.
- **Calibration:** Hair hygrometers are known to drift over time and exhibit hysteresis (the response to increasing humidity differs from the response to decreasing humidity). Therefore, periodic calibration against a more reliable humidity measuring instrument (like an Assmann psychrometer or a calibrated electronic hygrometer) is essential. Calibration involves:
  - Comparing the hygrograph's reading with the standard instrument's reading under various humidity conditions.
  - Adjusting the linkage mechanism or the pen position according to the manufacturer's instructions to align the hygrogram with the standard.
- **Cleaning the Instrument:** Gently clean the exterior of the hygrograph, ensuring that airflow to the hair bundle is not obstructed.
- **Hair Maintenance (Less Frequent):** Over time, the hairs can become contaminated with dust or oils, which can affect their responsiveness. Some manufacturers recommend occasional cleaning of the hair bundle, often with distilled water or a mild solvent, as per their instructions. Avoid touching the hairs directly with your fingers as oils from your skin can affect their performance.

By performing these maintenance tasks, especially regular calibration, you can ensure that the hair hygrograph provides a reasonably accurate continuous record of relative humidity. However, it's worth noting that hair hygrographs are generally less accurate than instruments like Assmann psychrometers, especially when high accuracy is required.

Useful Instruments

1. **Which instrument is used to measure air temperature?**
   a) Barometer b) Thermometer
   c) Hygrometer d) Anemometer
2. **What does a barometer measure?**
   a) Temperature b) Wind speed
   c) Atmospheric pressure d) Rainfall
3. **Anemometer is used for measuring:**
   a) Humidity b) Wind speed
   c) Solar radiation d) Atmospheric pressure
4. **Which instrument helps in determining relative humidity?**
   a) Thermometer b) Hygrometer
   c) Barometer d) Rain gauge
5. **Pyranometer is used to measure:**
   a) Air pressure b) Humidity
   c) Solar radiation d) Wind direction
6. **The instrument used to measure rainfall is:**
   a) Thermograph b) Evaporimeter
   c) Rain gauge d) Anemometer
7. **Which device measures the evaporation rate?**
   a) Barometer b) Thermograph
   c) Atmometer d) Evaporimeter
8. **A Stevenson screen is used to:**
   a) Protect meteorological instruments from direct sunlight b) Display weather forecasts
   c) Record rainfall d) Store marine data
9. **A wind vane is used to determine:**
   a) Wind speed b) Wind direction
   c) Wave height d) Air temperature

10. Which of the following is NOT a meteorological instrument?

a) Hygrometer
b) Pluviograph
c) Thermostat
d) Barometer

11. The maximum and minimum thermometer records:

a) Pressure variation
b) Rainfall changes
c) Daily temperature extremes
d) Wind direction changes

12. The term "automatic weather station" refers to:

a) A forecasting software
b) A manual observation station
c) A site with sensors recording weather data continuously
d) A radio station

13. Pluviograph differs from rain gauge because:

a) It is manual
b) It is more accurate
c) It provides continuous rainfall records
d) It measures wind speed

14. Which sensor is used for underwater temperature profiling in fisheries?

a) Float thermometer
b) Thermistor chain
c) Rain gauge
d) Barograph

15. Ceilometer is used to measure:

a) Sea surface temperature
b) Cloud height
c) Rainfall
d) Solar radiation

16. Why is temperature monitoring important in aquaculture?

a) Controls fish taste
b) Helps schedule feeding
c) Influences metabolic and growth rates
d) Reduces oxygen levels

17. Wind speed data is critical in fisheries for:

a) Improving breeding
b) Net design
c) Ensuring boat safety and predicting surface turbulence
d) Algae growth

18. Evaporation measurements are important for:

a) Fish marketing
b) Designing cages
c) Water budgeting in ponds
d) Fry selection

19. Which parameter is measured to assess storm risk in marine fisheries?

a) Humidity
b) Wind direction
c) Atmospheric pressure
d) Rainfall

20. Automatic Weather Stations are preferred because they:

a) Are costlier
b) Eliminate need for forecasts
c) Provide real-time, continuous data
d) Use manual recording

21. Use of solar radiation data in aquaculture helps to:

a) Predict fry movement
b) Enhance photosynthesis for plankton growth
c) Decrease pond depth
d) Increase water salinity

22. Cloud cover assessment helps in:

a) Net strength calculation
b) Predicting sunlight availability and evaporation
c) Reducing oxygen levels
d) Harvest timing

23. Relative humidity is important for aquaculture because:

a) It affects net elasticity
b) It influences evaporation and disease incidence
c) It affects fish age
d) It controls wave motion

24. Why is rainfall data essential in inland fisheries?

a) To schedule marketing
b) To estimate water availability and pond management
c) To feed fish better
d) To improve fish coloration

25. Thermistor chains are suitable in fisheries because:

a) They float
b) They monitor vertical water temperature changes
c) They record wave patterns
d) They help in harvesting

26. Which instrument is used to record wind speed over time?

a) Wind vane
b) Anemograph
c) Hygrograph
d) Pyrgeometer

27. What is the function of a pyrgeometer?

a) Measures air temperature
b) Measures net longwave radiation
c) Measures wave height
d) Measures rainfall

28. A barograph is a type of:

a) Rain gauge
b) Thermograph
c) Continuous atmospheric pressure recorder
d) Wind sensor

29. Which of the following is essential to monitor during a cyclonic alert for marine fisheries?

a) Temperature only
b) Pressure and wind speed
c) Solar radiation
d) Humidity alone

30. Which sensor type is crucial for automated rainfall measurement?

a) Float sensor
b) Tipping bucket
c) Thermal sensor
d) Anemometer

31. In coastal aquaculture, wind direction data is important for:

a) Determining feed quality
b) Predicting sediment resuspension and water quality shifts
c) Monitoring oxygen levels
d) Calculating harvest time

32. Which meteorological instrument helps decide safe fishing days?

a) Thermograph
b) Anemometer
c) Rain gauge
d) Wind vane

33. The data from a ceilometer is valuable in:

a) Predicting fish disease
b) Assessing cloud cover and flight visibility
c) Monitoring sediment levels
d) Estimating fish feed demand

34. Which device records temperature changes continuously?

a) Thermograph
b) Barometer
c) Wind vane
d) Hygrometer

35. What type of data does an AWS typically transmit automatically?

a) Only rainfall
b) Only wind speed
c) Multiple weather parameters including temperature, humidity, wind, and rainfall
d) Just atmospheric pressure

36. The function of a wave rider buoy in fisheries meteorology is to:

a) Track fish migration
b) Measure and transmit wave data
c) Measure rainfall
d) Record sunlight duration

37. In fish ponds, solar radiation is monitored to:

a) Decrease nutrient load
b) Support phytoplankton growth
c) Measure fish weight
d) Increase turbidity

38. A major benefit of installing AWS in coastal aquaculture farms is:

a) Improved net strength
b) Enhanced real-time weather monitoring and early warning
c) Reduced water salinity
d) Better boat design

39. Thermistors are commonly used in water for:

a) Wind speed measurement
b) Pressure fluctuation
c) Continuous temperature logging
d) Cloud tracking

40. Which of the following instruments is used in remote sensing meteorological satellites?

a) Barograph
b) Radiometer
c) Hygrometer
d) Wind vane

41. For surface water temperature, the best suited portable instrument is:

a) Float thermometer
b) Digital anemometer
c) Radiometer
d) Ceilometer

42. Why is rainfall monitoring crucial for inland fish seed farms?

a) To maintain taste
b) To prevent overflow or drying of nursery ponds
c) To calculate harvest time
d) To predict disease outbreaks

43. A pyranometer typically measures which solar radiation parameter?

a) Reflected solar radiation
b) Ultraviolet radiation
c) Total global solar radiation
d) Infrared radiation

44. The barometric pressure in marine zones helps in predicting:

a) Rain intensity
b) Wave direction
c) Storm development and sea level pressure drop
d) Salinity rise

45. Why are digital weather stations increasingly used in aquaculture?

a) To decorate farms
b) For real-time, multi-parameter data logging
c) Because they are more expensive
d) To monitor fish growth directly

46. An instrument commonly placed at 10 meters height for wind measurement is:

a) Rain gauge
b) Hygrometer
c) Cup anemometer
d) Thermistor

47. In the marine fisheries sector, atmospheric pressure is most useful to:

a) Detect oxygen level
b) Predict fish hatch
c) Track cyclone formation and landfall
d) Increase salinity

48. Which meteorological device is often mounted on buoys for offshore fisheries?

a) Barometer
b) Ceilometer
c) Automatic weather sensors
d) Rain gauge

49. Underwater temperature variations affecting fish behavior are best tracked using:

a) Barograph
b) Thermistor chain
c) Anemometer
d) Rain gauge

50. Integration of meteorological instruments with mobile technology benefits fisheries by:

a) Enhancing marketing
b) Reducing feed costs
c) Delivering real-time advisories and alerts
d) Increasing fish coloration

Answer Key (s)

1	b	2	c	3	b	4	b	5	c	6	c	7	d	8	a	9	b	10	c
11	c	12	c	13	c	14	b	15	b	16	c	17	c	18	c	19	c	20	c
21	b	22	b	23	b	24	b	25	b	26	b	27	b	28	c	29	b	30	b
31	b	32	b	33	b	34	a	35	c	36	b	37	b	38	b	39	c	40	b
41	a	42	b	43	c	44	c	45	b	46	c	47	c	48	c	49	b	50	c

15

Practical Schedule and Suggestions for Developing More Professional Outlook

The practical schedule to be implemented as per the syllabus prescribed for the course is reorganized in a logical sequence. The teachers offering the course is required to collect all the material required for effective practical training before the semester starts. In case of measurements made and observations taken, the students should be rigorous trained to understand significance of observation made and its practical utility (Fig. 15.1). The students are required to learn how to interpret and use the information contained in all products generated by meteorological services in fisheries and aquaculture.

- Graphical representation of structure of the atmosphere and physical layering,composition of the atmosphere
- Temperature instruments,Simple thermometers, Six's Maximum and Minimum Thermometer
- Preparation of isothermal maps,Interpretation of maps showing distribution of temperatures in the world and India
- Measurement of relative humidity, Dew point using Regnault's hygrometer, Dry and wet bulb thermometers and psychrometer
- Observation and identification of clouds
- 6.Measurement of rainfall using rain gauge, preparation of rainfall charts
- Preparation of isohyetal maps and interpretation of maps showing distribution of rainfall during southwest monsoon and northeast monsoon seasons
- Measurement of atmospheric pressure using Fortin's barometer and aneroid barometers ,preparation and interpretation of isobaric maps
- Measurement of wind direction using wind vane and wind speed using Robinson's cup anemometer
- Diagrammatic representation planetary pressure and wind systems
- Self recoding instruments Thermograph, Hygrograph and Barograph
- Interpretation of satellite weather pictures

- Interpretation of weather forecasts in local and regional context
- Diagrammatic representation of shape,size,structure,zones.
- Latitudes, Longitudes, great circles
- Typical landscape mapping,map reading
- Geographical terms used in land scape

Fig. 15.1: Professional outlook in fisheries and acquaculture

15.1 Suggestions for Skill Development, Professionalism and Competence to Address Real World Problems

To make the practical meteorological training for fisheries and aquaculture students more engaging and effective, and to instill a professional outlook, faculty can adopt a multi-faceted approach. Here's a breakdown of key areas and specific strategies:

15.1.1 Creating Interest in Learning

• Contextualization

- **Real-World Case Studies:** Start each practical session with a real-world case study of how meteorological conditions impacted fisheries or aquaculture operations. This could include examples of:
  - Storm surges affecting coastal aquaculture farms.
  - Temperature changes influencing fish migration patterns.
  - Wind patterns affecting fishing vessel safety.
  - El Niño/La Niña events impacting global fish stocks.
- **Connect to Local Relevance:** Emphasize how weather patterns in the students' specific geographic location affect local fisheries and aquaculture practices. This makes the subject matter more relatable and important to them.

- **Guest Speakers:** Invite professionals from the fisheries and aquaculture industry, as well as meteorologists specializing in marine or coastal environments, to share their experiences and insights. This can expose students to the practical applications of meteorological knowledge.

- **Active Learning**
  - **Hands-on Activities:** Design practical sessions that go beyond simply taking measurements. Incorporate activities that require students to:
    - Analyze weather maps and predict their impact on fishing operations.
    - Use meteorological instruments to collect data and then make decisions based on that data (e.g., simulating a scenario where they must advise a fishing fleet on whether it's safe to go out).
    - Develop simple models to illustrate the relationship between weather variables and aquaculture productivity.
  - **Problem-Based Learning:** Present students with realistic problems faced by the fisheries and aquaculture industry that require them to apply their meteorological knowledge. For example:
    - "A harmful algal bloom is developing in a coastal area. Analyze the weather conditions and predict its potential spread and impact on local fish farms."
    - "A fishing vessel is caught in a storm. Use meteorological data to determine the safest course of action."
  - **Collaborative Projects:** Encourage students to work in teams on practical projects that simulate real-world scenarios. This fosters teamwork, communication, and problem-solving skills.

- **Technology Integration:**
  - **Interactive Simulations:** Use weather simulation software and online resources to provide students with dynamic and interactive learning experiences.
  - **Data Visualization Tools:** Introduce students to tools for visualizing meteorological data, such as plotting software and GIS, to help them understand complex weather patterns.
  - **Mobile Apps:** Explore the use of weather apps .and other mobile technologies that are relevant to fisheries and aquaculture operations.

15.1.2 Skill Development

- **Measurement and Observation Skills**
  - **Instrument Proficiency:** Provide thorough training on the use of various meteorological instruments, emphasizing accuracy, precision, and error analysis.
  - **Systematic Observation:** Teach students how to make systematic and detailed observations of weather phenomena, including cloud types, wind direction and speed, and sea state.
  - **Data Recording and Management:** Train students in proper techniques for recording, organizing, and managing meteorological data.
- **Analytical and Critical Thinking Skills**
  - **Data Interpretation:** Develop students' ability to interpret meteorological data, identify trends, and draw meaningful conclusions.
  - **Forecasting Skills:** Provide opportunities for students to practice making short-term weather forecasts, and to evaluate the accuracy of their predictions.
  - **Risk Assessment:** Train students to assess the risks associated with different weather conditions and to make informed decisions to mitigate those risks.
- **Communication Skills**
  - **Technical Writing:** Require students to write clear and concise reports on their observations and analyses (Fig. 15.2).
  - **Oral Presentations:** Provide opportunities for students to present their findings to the class, using appropriate visual aids.
  - **Communication with Stakeholders:** Simulate scenarios where students must communicate meteorological information to non-technical audiences, such as fishers or aquaculture farmers.
- **Information Literacy**
  - **Sourcing Reliable Information:** Teach students how to identify and access reliable sources of meteorological information, including government agencies, research institutions, and reputable weather services.
  - **Evaluating Information:** Train students to critically evaluate the quality and relevance of meteorological information from different sources.

- **Using Meteorological Products:** Provide in-depth instruction on how to interpret and use various meteorological products, such as:
  - Weather maps (surface analysis, upper-air charts).
  - Satellite imagery.
  - Radar data.
  - Numerical weather prediction models.
  - Marine forecasts.
  - Tide tables.

Fig. 15.2: Skill development in fisheries

15.1.3 Fostering a Professional Outlook

- **Ethical Considerations:** Discuss the ethical responsibilities of professionals who use meteorological information in the fisheries and aquaculture industry, such as providing accurate and timely warnings of hazardous weather conditions.
- **Professional Standards:** Introduce students to the professional standards and best practices in the field of meteorology.
- **Lifelong Learning:** Encourage students to develop a mindset of continuous learning and professional development, emphasizing the importance of staying up-to-date with the latest advances in meteorological science and technology.

- **Mentorship**
  - Provide students with opportunities to interact with professionals in the field, such as through guest lectures, workshops, or mentoring programs.
  - Offer guidance and support to students as they explore career options and prepare for the transition from academia to the workplace.

- **Industry Engagement**
  - Facilitate internships or field placements for students with fisheries and aquaculture companies or organizations.
  - Encourage students to attend professional conferences and workshops related to meteorology, fisheries, and aquaculture.

By implementing these strategies, faculty can play a crucial role in creating a dynamic and engaging learning environment that not only equips students with the necessary meteorological knowledge and skills but also fosters a professional outlook and prepares them for successful careers in the fisheries and aquaculture industry.

Practical Schedule and Suggestions for Developing more Professional Outlook

1. **Which of the following is essential in a practical schedule for fisheries meteorology?**
   a) Marketing fish
   b) Observing cloud types
   c) Feeding schedules
   d) Measuring fish weight
2. **What is the ideal duration for basic meteorological field training in a semester?**
   a) 2 days
   b) 1 week
   c) 2–3 weeks
   d) Full semester
3. **Recording atmospheric pressure is important because it helps:**
   a) Identify fish feed type
   b) Predict storms and wind patterns
   c) Calculate salinity
   d) Reduce pollution
4. **A practical exercise in geography should include:**
   a) Pond harvesting
   b) Contour mapping and map reading
   c) Water coloring
   d) GPS for aquaculture location only
5. **Which instrument should be demonstrated in field classes for wind measurement?**
   a) Thermometer
   b) Anemometer
   c) Rain gauge
   d) Compass
6. **One goal of practical training in meteorology for fisheries students is to:**
   a) Reduce feed cost
   b) Enhance awareness of weather's impact on aquatic life
   c) Promote urban fishing
   d) Study marine pollution only

7. A standard practical schedule should focus on:

a) Weekly fish market visits
b) Monthly swimming practice
c) Hands-on instrument use and data interpretation
d) Literature review only

8. A major component of geographic skills for fisheries includes:

a) Net weaving
b) Cartographic representation of water bodies
c) Daily temperature logging
d) Counting aquatic insects

9. The integration of meteorology into fisheries training ensures:

a) Automatic fish production
b) Seasonal forecasting skills for resource planning
c) More feeding hours
d) Shorter harvest periods

10. GPS training in geography helps in:

a) Deep water fishing
b) Mapping and monitoring aquatic sites
c) Sorting fish varieties
d) Raising pond temperature

11. What can increase professional outlook among fisheries students?

a) Frequent exams
b) Structured exposure to field-based instruments
c) Indoor-only activities
d) Theory-based lectures only

12. Understanding climate variability is important because it:

a) Stops disease
b) Helps in fish packaging
c) Affects fish breeding and migration
d) Reduces salt in water

13. One suggested improvement in practical training is:

a) Removing field work
b) Increasing classroom hours only
c) Including weather report interpretation and marine forecast reading
d) Visiting zoos

14. A professional in fisheries meteorology should:

a) Avoid digital tools
b) Only work in the lab
c) Correlate meteorological data with aquatic species' behavior
d) Study forest species

15. A student-friendly practical schedule should be:

a) Flexible and application-oriented
b) Theoretical only
c) Focused on exams
d) Limited to lecture slides

16. Professional development includes:

a) Ignoring weather forecasts
b) Using mobile apps for meteorological data
c) Focusing only on pond fish
d) Avoiding climatic discussions

17. Weather journals/logbooks are maintained to:

a) Fill time
b) Track daily climatic conditions for learning patterns
c) Impress teachers
d) Avoid practicals

18. Inclusion of GIS in training helps students:

a) Find fish markets
b) Analyze spatial data for aquaculture zones
c) Train in fish behavior
d) Count rainfall

19. Regular use of weather instruments builds:

a) Artistic sense
b) Habit of forecasting and pattern recognition
c) Laziness
d) Marine interest only

20. Practical sessions should be conducted:

a) Only in winter
b) Indoors only
c) During different seasons for diverse exposure
d) Monthly once

21. Participation in real-time data analysis:

a) Confuses students
b) Builds problem-solving skills
c) Delays learning
d) Is irrelevant

22. Weather forecasting training is valuable for:

a) Decoration
b) Aquatic risk management
c) Classroom ranking
d) Chemistry labs

23. What boosts student interest in practicals?

a) More theory books
b) Group field activities and local weather analysis
c) Sitting exams weekly
d) Watching weather TV only

24. Why integrate professional talks into training?

a) For break time
b) To inspire with real-life experiences
c) To fill schedule
d) For external marks

25. What should follow every field activity?

a) Free time
b) Formal feedback and learning session
c) Party
d) Online quiz

26. A practical schedule in meteorology should aim to:

a) Increase net-making skills
b) Develop analytical thinking through data collection
c) Focus on cooking fish
d) Avoid use of field stations

27. Which of the following enhances observational skills in weather monitoring?

a) Watching documentaries
b) Field observation with data logging
c) Reading outdated textbooks
d) Group discussions only

28. Professional training must integrate:

a) Fancy PowerPoint shows
b) Real-time monitoring tools like AWS, radar, or satellite images
c) Monthly grading
d) Storekeeping lessons

29. A basic map-reading session in geography enhances:

a) Map coloring ability
b) Aquatic product processing
c) Spatial orientation and resource planning
d) Interest in weather myths

30. Exposure to local weather stations helps students:

a) Learn from farmers only
b) Understand real-world application of classroom concepts
c) Get out of exams
d) Learn about soil texture

31. What is a key element of weather-related decision-making in fisheries?

a) Price forecasting
b) Wind and temperature forecasts
c) Boat color
d) Distance to market

32. Which practice supports a stronger professional identity in students?

a) Maintaining personal weather diaries
b) Memorizing weather codes
c) Avoiding group tasks
d) Avoiding feedback

33. Which mobile app is widely used for real-time weather updates?

a) Google Maps
b) IMD Mausam
c) MS Paint
d) Aqua Tracker

34. Collaborative learning in field exercises promotes:

a) Individual competition
b) Teamwork and professional behavior
c) Passive observation
d) Time wasting

35. One way to assess field-based learning is through:

a) Oral storytelling
b) Written reflections and practical reports
c) Verbal chats only
d) Avoiding documentation

36. Regular exposure to synoptic charts develops:

a) Communication skills
b) Interpretation ability of large-scale weather patterns
c) Fishing net tying
d) Field safety rules

37. What increases confidence among students in applying met knowledge?

a) Attending cultural events
b) Interacting with meteorological professionals
c) Watching weather fiction
d) Guessing forecasts

38. One benefit of learning GIS in fisheries geography is:

a) Calculating fish feed
b) Locating and managing aquaculture zones
c) Net mending
d) Estimating water temperature

39. Long-term skill development is supported by:

a) One-time lecture
b) Regular hands-on sessions and practice
c) Watching recorded sessions
d) Ignoring weather forecasts

Answer: b) Regular hands-on sessions and practice

40. A weather forecast bulletin is most useful when:

a) It's memorized
b) It's analyzed for impact on fish habitats
c) It's ignored
d) It is printed only

41. The use of logbooks in fieldwork supports:

a) Routine writing
b) Development of scientific documentation and tracking skills
c) Reading habit
d) Literature reviews

42. One way to promote professional behavior during training:

a) Ignore time schedules
b) Practice punctuality and structured planning
c) Work alone
d) Use only theoretical content

43. Digital weather boards on campus can:

a) Entertain students
b) Encourage timely weather awareness and updates
c) Replace mobile phones
d) Be ignored

44. Participation in student-led forecast presentations fosters:

a) Public speaking and analytical skills b) Unnecessary pressure
c) Passive learning d) Avoiding weather terms

45. What motivates students during field visits?

a) Free food b) Role in real-time data collection and observation
c) Long breaks d) Group photos only

46. Weekly weather summary preparation helps in:

a) Enhancing data analysis and reporting skills b) Reducing classroom time
c) Playing weather games d) Submitting notes faster

47. Real-time analysis of rainfall data enables:

a) Tracking weather patterns relevant to aquaculture b) Avoiding classes
c) Designing fish tanks d) Creating maps manually

48. Suggested improvement in curriculum:

a) Remove practicals b) Add real-time and location-based weather data use
c) Increase theory hours d) Focus on cooking methods

49. How can digital platforms support student learning in meteorology?

a) Watching random videos b) Accessing satellite data and cloud pattern updates
c) Avoiding books d) Using text-only methods

50. Which activity marks a professionally engaging classroom?

a) Memorizing facts b) Applying knowledge to real-world weather scenarios in fisheries
c) Ignoring forecasts d) Only textbook reading

Answer Key (s)

1	b	2	c	3	b	4	b	5	b	6	b	7	c	8	b	9	b	10	b
11	b	12	c	13	c	14	c	15	a	16	b	17	b	18	b	19	b	20	c
21	b	22	b	23	b	24	b	25	b	26	b	27	b	28	b	29	c	30	b
31	b	32	a	33	b	34	b	35	b	36	b	37	b	38	b	39	b	40	b
41	b	42	b	43	b	44	a	45	b	46	a	47	a	48	b	49	b	50	b

Bibliography

Books

Mavi, HS. 1986. Introduction to Agrometeorology. Oxford & IBH Publishing Co Pvt.Ltd. ISBN: 978-8120409101. Pp 296.

Murthy, V.R.K. 2015. Basic Principles of Agricultural Meteorology. BSP Books Private Limited. ISBN: 978-9352300747. Pp 288.

Prasada Rao, G.S.L.H.V. 2008. Agricultural Meteorology. PHI Learning Pvt Ltd. ISBN: 978-8120333383. Pp 384.

Ramana Rao, B.V. and V. Uma Maheswara Rao. 2024. Agricultural Meteorology. Jain Brothers. ISBN: 978-9390576975. Pp 154.

Ramana Rao B.V., Surender Singh, V. Uma Maheswara Rao. 2025. Preparatory Microclimatology. Publisher: Professional Prints. New York USA. ISBN: 978-1966695189. Pp 167.

Ramana Rao B.V., Surender Singh, V. Uma Maheswara Rao. 2025. Conceptual Meteorology. Publisher: New India Publishing Agency, New Delhi. ISBN 978-9358874501. Pp 152.

Ramana Rao B.V., Surender Singh, V. Uma Maheswara Rao. 2025. Primer on Agrohydrometeorology. Publisher: Professional Prints. New York USA. ISBN: 978-1966695301. Pp 144.

Ramana Rao B.V., Surender Singh, V. Uma Maheswara Rao, Anu Bulusu. 2025. Farming Weather: Harnessing Citizen Science for Smart Agriculture. Publisher: Professional Prints. New York USA. ISBN: 978-1966695561. Pp 154.

Ramana Rao B.V., Surender Singh, V. Uma Maheswara Rao. 2025. Forecasting the Future: Effective Climate Guidance on Monsoons for Farm Prosperity. Publisher: Professional Prints. New York USA. ISBN: 978-1966695912. Pp 246.

Ramana Rao B.V., Surender Singh, V. Uma Maheswara Rao. 2025. Operational Agricultural Meteorology: A Realm for Sustainable Livelihoods. Publisher: Professional Prints. New York USA. ISBN: 978-1966695943. Pp 245.

Ramana Rao, B.V., Surender Singh, V. Uma Maheswara Rao. 2025. System Simulation and Agromet Advisory Services. Publisher: New India Publishing Agency, New Delhi. ISBN: 978-9372196191. Pp 510.

Ramana Rao, B.V., Surender Singh, V. Uma Maheswara Rao, Anu Bulusu. 2025. Climate Crisis and Farming Intelligence. Publisher: Professional Prints. New York USA. ISBN: 978-1966695646. Pp 159.

Ramana Rao, B.V., Surender Singh. 2025. Advancing Agricultural Research: Journal Rating, Leadership and Challenges. Publisher: Professional Prints. New York USA. ISBN: 978-1966695-608. Pp 189.

Ramana Rao, B.V., Surender Singh, V. Uma Maheswara Rao, Anu Bulusu. 2025. Weather Intelligence and Analytics: Response Management in Agriculture. Publisher: Professional Prints. New York USA. ISBN: 978-1966695523. Pp 217.

Seemann, J.; Chirkov, Y.I.; Lomas, J. and Primault, B. 1979. Agrometeorology. Springer Berlin, Heidelberg. ISBN: 978-3642672903. Pp 326.

Subramaniam, H. 2019. Marine Meteorology: Publisher: Vijaya Publications. ASIN B082XB2666. Pp 308.

Surender Singh., V. Uma Maheswara Rao, B.V. Ramana Rao. 2025. Weather Forecasting. Publisher: New India Publishing Agency, New Delhi. ISBN: 978-9358875492. Pp 392.

Surender Singh., V. Uma Maheswara Rao, B.V. Ramana Rao. 2025. Introductory Agricultural Meteorology. Publisher: New India Publishing Agency, New Delhi. ISBN: 978-9358871487. Pp 412.

Surender Singh. 2025. MCQs on Agricultural Meteorology. Publisher: New India Publishing Agency, New Delhi. ISBN: 978- 9372196450. Pp 339.

Varshneya, M.C. and Pillai, B.P. 2004. Textbook of Agricultural Meteorology. ICAR Publication. ISBN: 978-8171640195. Pages 242.

Journals

Agricultural and Forest Meteorology.
https://www.sciencedirect.com/journal/agricultural-and-forest-meteorology
Chinese Journal of Agrometeorology.
https://zgnyqx.ieda.org.cn/EN/1000-6362/home.shtml
Italian Journal of Agrometeorology.
https://riviste.fupress.net/index.php/IJAm/issue/view/15
Journal of Agril Meteorology.
:https://journals.indexcopernicus.com/search/details?id=46231
Journal of Agrometeorology.
https://journal.agrimetassociation.org/index.php/jam/issue/view/73
MAUSAM.
https://mausamjournal.imd.gov.in/index.php/MAUSAM
Marine Meteorology
https://www.sciencedirect.com/topics/earth-and-planetary-sciences/marine-meteorology

Web Resources

https://www.cmfri.org.in/
https://www.cift.res.in/
https://mausam.imd.gov.in/
https://imdagrimet.gov.in/
https://www.tropmet.res.in/
https://nwp.ncmrwf.gov.in/
https://www.skymetweather.com/
https://icar.org.in/
https://www.icar-crida.res.in/
https://niasm.icar.gov.in/
https://www.sac.gov.in/Vyom/
https://www.isro.gov.in/
https://www.icrisat.org/
https://www.ecmwf.int/
https://wmo.int/
https://www.fao.org/climate-smart-agriculture/en/
https://www.agrimetassociation.org/

https://safoam.org.in/
https://www.insam.org.in/
https://www.lidsen.com/journals/aeer/aeer-special-issues/Air-Sea-Interact-Marin-Meteorolog

Subject Index